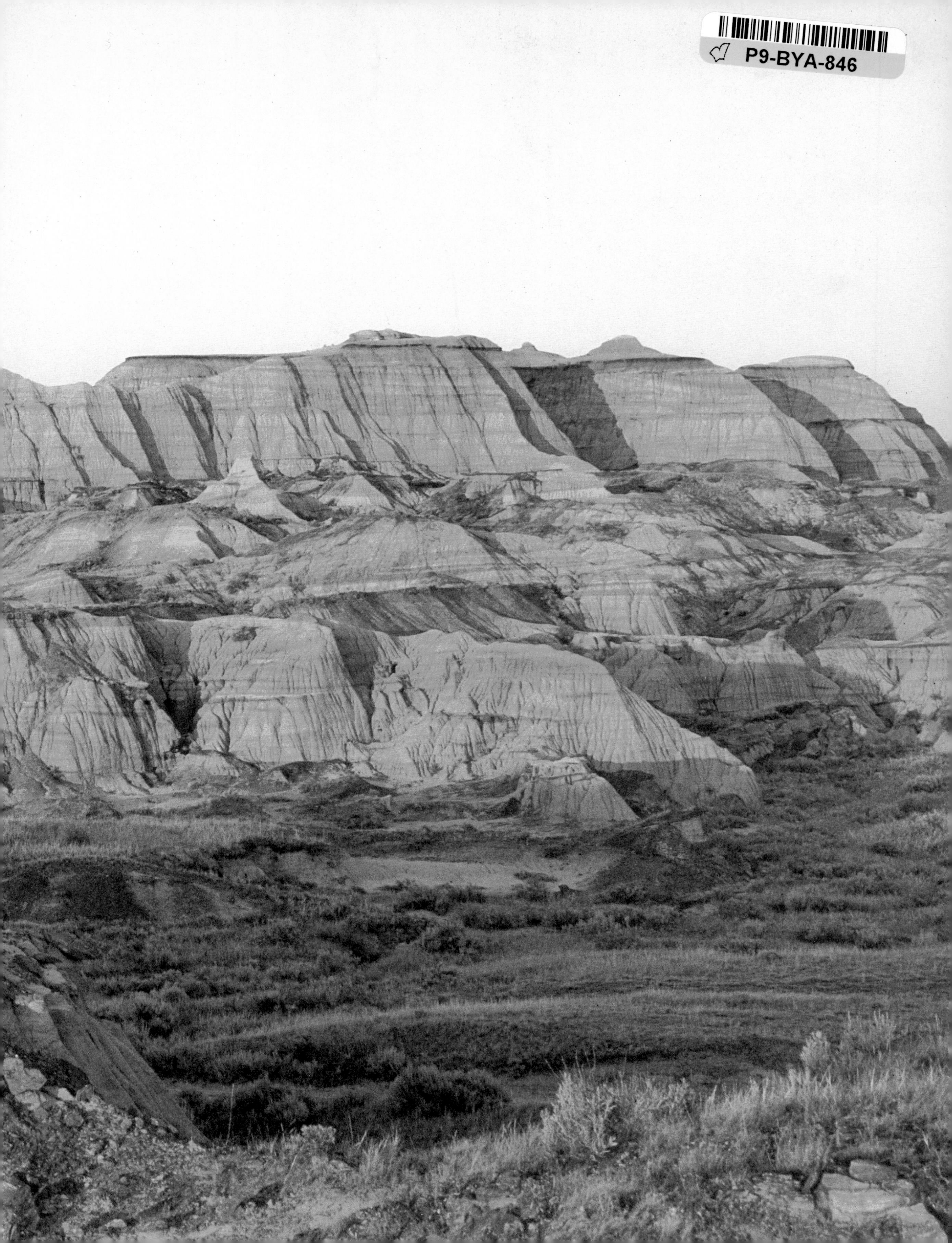

THE
MIRACLE
PLANET

THE
MIRACLE
PLANET

BRUCE BROWN AND LANE MORGAN

GALLERY BOOKS
An Imprint of W. H. Smith Publishers Inc.
112 Madison Avenue
New York City 10016

SCIENTIFIC ADVISERS

DR STEPHEN C. PORTER
Professor of Geological Science, University of Washington
Director, Quaternary Research Center

DR BRIAN J. SKINNER
Eugene Higgins Professor of Geology and Geophysics Yale
University Chair, National Academy of Sciences Board on
Earth Sciences

DR DENNIS L. HARTMANN
Professor, Department of Atmospheric Sciences,
University of Washington

This edition published in 1990 by Gallery Books,
an imprint of W.H. Smith Publishers, Inc.,
112 Madison Avenue, New York, New York 10016

Gallery Books are available for bulk purchase for sales
promotions and premium use. For details write or telephone
the Manager of Special Sales, W.H. Smith Publishers, Inc.,
112 Madison Avenue, New York, New York 10016. (212) 532-6600

Produced by Weldon Owen Pty Limited
43 Victoria Street, McMahons Point NSW 2060, Australia
Telex AA23038, Fax (02) 929 8352
A member of the Weldon International Group of Companies
Sydney · Hong Kong · London · Chicago · San Francisco.

Publishers: Kevin Weldon, John Owen
Publishing Manager: Stuart Laurence
Managing Editor: Kim Anderson
Editor: Christine Mackinnon
Copy Editor: Camilla Sandell
Index: Barbara Crighton
Maps: Greg Campbell
Designer: Kathie Baxter Smith
Production Director: Mick Bagnato
US Consultant Editors: Barry Stoner, Katie Jennings, Simon Griffith

In association with KCTS Television
401 Mercer Street, Seattle WA 98109, USA
President and CEO: Burnill F. Clark
For The Miracle Planet television series:
Executive Producer: Barry Stoner
Associate Producer: Katie Jennings
Research Co-ordinator: Simon Griffith

© Text 1990 Weldon Owen Pty Limited
© Text 1990 Weldon Owen Inc.
© Photographs 1990 Japan Broadcasting Corporation and
Japan Broadcast Publishing Co Ltd

World publishing rights arranged with
Japan Broadcasting Corporation through Japan UNI Agency,
Inc, Tokyo.

ISBN 0-8317-5999-2

Typeset by Keyset Phototype
Printed by Griffin Press Limited, Marion Road, Netley, South Australia

A Weldon Owen Production

CONTENTS

INTRODUCTION

Throughout history, humans have wondered about the planet Earth: How was it formed? What is our place upon it? What will the future bring? The legends we have created to explain these mysteries underscore the power they exert over our imaginations.

In recent years, scientists have come closer to the central riddles of creation. Their journeys above the Earth and beneath the sea rival those of the legendary heroes. And some of their conclusions sound as fantastic as the tales told around ancient campfires. The more we learn, the more the development of the Earth and life upon it seems no less than miraculous.

Although all the planets of our solar system were created at the same period out of the same materials, only Earth is known to support life. Only Earth has the combination of solid ground, a sheltering atmosphere, and nourishing waters. Only on Earth have beings evolved who are capable of consciously destroying this complex and beautiful system.

Beginning with the formation of the solar system, *The Miracle Planet* explores characteristics that have combined to make Earth unique. It shows the tremendous forces and vast expanses of time underlying familiar sights. The geysers of Yellowstone National Park are the result of the same processes that have rearranged continents and pushed sea floors miles into the air. The white cliffs of Dover and the eerie mountains of Chinese landscapes are more than cultural icons, they are a visible part of a chemical cycle that maintains the equilibrium of the atmosphere. The bacteria that help cows convert grass into beefsteak are in hiding, refugees from a world where life abounded in the absence of oxygen. The vast grainfields of the American heartland are more than a symbol of enterprise and productivity, they are a sign of human influence — for good or ill — on the biosphere that supports us.

A companion to the television series produced by KCTS Television and NHK/Japan, *The Miracle Planet* is more than a general survey of our current store of knowledge. By introducing competing theories and historical context, it provides insight into how science works. There is no machine-like progression from one discovery to the next. Knowledge advances by fits and starts, and intuition and luck share honours with painstaking research. Even when dealing with events billions of years before our own beginnings, there is always a human factor.

How we will live and whether we will flourish in the future depends on how well we understand our world. *The Miracle Planet* has been written to help us understand it.

◄ At the birth of our solar system, the remains of an exploded star are beginning to reconstitute into a swirling mass of incandescent gases orbited by rocky fragments.

►► In the course of a "demolition derby of cosmic proportions," some planetesimals gained enough mass to withstand future collisions. The largest of these became the planets of our solar system.

THE THIRD PLANET

▲ A telescope near the summit of Mauna Kea, Hawaii. Its location means that astronomers can enjoy the advantages of high altitude without arctic weather.

HUMAN beings have long stared with wonder at the stars and populated the heavens with creatures of their own imagination.

The ancient Greeks looked up at the Pleiades and saw the seven daughters of Atlas, transformed into stars to escape the unwanted attentions of Orion, the giant hunter; 5,000 miles away, the Cherokee saw a group of boys who danced into the sky rather than obey their mothers' orders.

Stargazers today may view the same constellation from a different perspective. They may gauge the degree of air pollution by the number of stars they can see. They may know that the seven sisters (or six delinquents) are part of a cluster of several hundred young stars 400 light years away. They may be able to describe the intensely hot thermonuclear reactions that produce the icy light overhead.

Once a potent source of myth and parable, the stars and planets have now moved into the world of complex equations and computer imagery. The legends of the constellations may have lost some of their power in the process, but the wonder of the skies is undimmed. The earth is linked to its galactic neighbors — stars, comets, and fellow planetary wanderers — in ways that the ancient stargazers never dreamed.

Our very existence is due, in part, to the meteorites and comets that collided with a rocky aggregation of debris in the early years of our solar system. All of the crucial chemical elements of a habitable planet, and all of the basic compounds in living things, have since been found in meteorites. As far as we know, the whole universe is seeded with the elements of life. Each shooting star is a message from the rest of space, an invitation to look beyond ourselves in order to understand our common origins.

In order to respond to that invitation, modern observers must find ways to evade the distortions of our protective atmosphere. Pioneers of the telescope mounted their equipment on the rooftops of downtown Greenwich or Danzig, but modern scientists must look further afield. One favored spot is Mauna Kea, which rises nearly 13,800 feet above the beaches on the island of Hawaii. From the dormant volcanic craters on its

▲ For astronomers, the best views from Mauna Kea come after sundown. The telescopes are above the island's tropical inversion layer, which traps much of the moisture and pollutants that complicate astronomy.

▶ The shock waves of a supernova explosion course through space, agitating and compressing clouds of interstellar gas. The pull of gravity will cause the larger clumps to contract and spin. Gravitational energy is converted to heat as the contraction progresses, and if the heat is intense enough, a nuclear reaction begins at the core of the gaseous mass. The Sun had enough mass to follow this scenario; the planet Jupiter, the next largest body in our solar system, did not.

summit, one can look down on the rainforest clouds below and look up into distant galaxies. Mauna Kea's dry air, atmospheric stability, and relative freedom from light and air pollution make it a favored astronomical site, home of what astronomers call "good seeing." The observatory near its summit houses telescopes from several countries.

Astronomers at Mauna Kea train their instruments on the solar system and beyond. From Earth's vantage point at the edge of the Milky Way, submillimeter telescopes record invisible radiation from dense patches of gas in the center of the galaxy. These coalescences are the birthplaces of new stars. Other telescopes collect light that began its journey from a distant galaxy before the human species evolved.

Astronomers also study our neighboring planets, their satellites, and the asteroids ringed between Mars and Jupiter. Planetary atmospheres and volcanic activity are of particular interest, since they may elucidate our common origins and our divergent paths. From inhospitable Venus to welcoming Earth, all the planets, in fact all the universe, were formed from the same basic constituents. But from these common components has evolved an astonishing diversity.

VIOLENT BEGINNINGS

The planets were once thought of as varicolored versions of our own, populated by beings created more or less in our own image. In fact they are strange worlds, and their "little green men" and "bug-eyed monsters" are creatures of our self-centered imaginations. The stars, wheeling

impassively above us, turn out on closer inspection to be maelstroms, wracked by explosions and subject to bizarre forces at the limits of space and time. In one corner of this infinity of commotion, about 4.6 billion years ago, our own home planet was formed.

No one can say precisely what happened at that time, but scientists have constructed a plausible model for the creation of the solar system. Dust

and gases, probably from an exploded star, coalesced into a huge, whirling disk. At its center, where gravitational pull was strongest, the compressed and superheated gases started thermonuclear reactions that converted hydrogen into helium, producing intense energy. This radiant mass was the primitive Sun.

In the environs of the protosun, dust specks collided and combined, eventually forming trillions of rocky fragments. Collisions continued, sometimes disintegrating the planetesimals (or very early meteorites) back into dust, sometimes combining them into larger aggregates. As the larger bodies grew in mass, their increasing gravity attracted still more planetesimals. The result was what science writer Marcia Bartusiak calls "a demolition derby of cosmic proportions."

In the universe's time frame, it did not take long to establish the winners. Within a few hundred million years of our Sun's formation, each planet in the solar system had established an orbital path and had almost reached its present size. In the space between Mars and Jupiter, a ring of planetesimals that had failed to coalesce formed the asteroid belt.

A sign of the planets' violent beginnings is the impact crater. Colliding meteorites and comets have left their scars throughout the solar system. Although Earth was also formed and marked by this process, its subsequent evolution has destroyed much of the evidence. Scientists received their first indications of the process, not from craters in their own landscapes, but by looking at the Moon. Galileo Galilei published a description and sketches of lunar mountains and craters in 1610, based on observations with one of the first telescopes. Galileo concentrated on the mountains, estimating their height by the angle of their shadows. As early as 1826, the Bavarian astronomer Franz von Paula Gruithuizen proposed that lunar craters were the result of meteorite impacts. He was right, but his credibility was undermined by his previous assertions that other lunar features were built by a race of moon creatures he called Selenites.

As more sophisticated telescopes and less eccentric scientists turned to the Moon, the meteorite impact theory began to gain adherents. In the late 1960s, rock samples brought back to Earth from Moon landings disproved alternative explanations — such as volcanic eruptions — for its craters. Some 500,000 lunar craters are visible from Earth, and in the last two decades, space

▲ The raw material for the creation of the solar system probably came from a supernova, the explosion of a star more than ten times the mass of the Sun. Supernovae send quantities of carbon, silicon, gold, lead, and uranium into space, as well as more common universal elements such as hydrogen and helium.

◄ A nebula like this one may be formed of debris from a stellar explosion or cast-off gas from a red giant star. The visible light of many nebulae is faint and diffuse, but radio and X-ray telescopes pick up strong signals from the intensely radioactive gases.

TELESCOPES

WHEN Galileo Galilei trained his telescope on the circling moons of Jupiter in 1609, he also brought into focus the burgeoning theories of heliocentrism. For if other planets had moons, just as Earth does, Earth might be one among many and not the center of the universe. It might in fact be part of a solar system, with a star at its center. The idea was far from new — Aristarchus of Samos had broached it nearly two thousand years before — but it took visible proof to back up theoretical conclusions.

The early telescopes made it clear that the universe was bigger and more crowded than unaided vision could encompass. Once that boundary of observation was breached, stargazers have accepted no further limits. Human eyesight has grown no keener since Galileo, but our machines can translate the invisible universe into messages we can comprehend.

Optical telescopes like Galileo's magnify the light transmitted in the visible spectrum, making small objects seem larger and faint ones brighter. Their successors, immensely larger and more sophisticated, equipped with cameras, video screens, and computers to record and analyze the images they find, are still part of the astronomers' arsenal. But many of the discoveries of the last century come from messages transparent to the human eye.

Whether it comes from a lamp or the star Alpha Centauri, all visible light travels in the same narrow band of wavelengths. Even that restricted range can crowd a moonless night, but it is like one station on a crowded dial, comprising only about 10 percent of the radiation broadcast in the universe. Invisible radiation ranges on both sides of the visible spectrum, from the large and long — radio waves averaging up to six miles from peak to peak and a thousand feet from top to bottom — to the smallest, fastest-pulsing gamma rays, no larger than an atomic nucleus. Most wavelengths on the electromagnetic spectrum are blocked by the atmosphere and never reach the Earth. Radiation on the low-frequency end of the spectrum, such as radio waves, penetrates the atmosphere with ease but is simply outside our retina's capacity.

Radio astronomy began in 1931 when Karl Jansky, a young radio engineer for Bell Telephone Laboratories, began seeking the source of static that interfered with transatlantic telephone communication. Jansky concluded that the annoying hiss was "star noise," broad-spectrum radiation from the center of the Milky Way. (His discovery is commemorated in the term jansky, a unit used to measure the strength of an incoming electromagnetic wave signal.)

Today, radio telescopes — linked networks of latticework parabolas resembling home satellite dishes — pick up signals from the clouds of cosmic debris surrounding the birthplace of stars. They also record the "signature" wavelengths of free-floating atoms and molecules in interstellar space. These chemicals — hydrogen, formaldehyde, methyl alcohol, and others — represent the basic components of the universe. Even planets can broadcast on the radio spectrum, as Jupiter does.

In 1965, two more Bell Telephone scientists made another epochal and accidental discovery. Arno A. Penzias and Robert Wilson were reworking a telescope to study microwave emissions from beyond the Milky Way when they were sidetracked by an unexplained background signal. The interference was minor — for a while it was attributed to pigeon droppings on the telescope — but Penzias and Wilson were determined to find its source.

After a year, their difficulties came to the attention of Robert Dicke, a Princeton astronomer who was looking for trace radiation from the Big Bang. He realized that Penzias and Wilson had found it. The heat from the fireball of the universe's birth had dissipated, but it was not gone. As a piece of burning wood cools through the spectrum from yellow-white flames to reddish embers to black coals, radiating heat in ever-lengthening waves, the heat of the Big Bang had diminished to about 2.7 degrees Kelvin, just above absolute zero (which is -450 degrees Fahrenheit or -273 degrees Celsius), broadcasting in the microwave range. It was this heat, more than 10 billion years old, that Penzias and Wilson had picked up. In 1978 they were awarded the Nobel Prize in physics.

Radio astronomy also helped confirm the existence of one of the strangest stellar entities. Jocelyn Bell-Burnell was a Cambridge graduate student in 1967 when she noted a quarter-inch of anomalous transmission in the miles of charts produced by a huge radio telescope. It recorded a pulsing transmitter whose speed and regularity fit no known cosmic source. Discomfited Cambridge researchers had to consider that it might be a message from intelligent beings.

Then Bell turned up similar sources in other parts of the sky. Since it seemed beyond coincidence that several civilizations light years apart would choose the same form of contact, Bell and her co-workers abandoned the designation LGM (for Little Green Men) and concentrated on an inanimate source. They settled on the neutron star, which had been predicted since the 1930s but had never been observed. Neutron stars stretch the boundaries of physical laws, packing a mass greater than the Sun into a diameter of only a few miles. This incredible density allows extremely rapid rotation and a strong magnetic field, turning the star into an electrical generator.

Observations on the faster end of the radiation spectrum depend on getting up above the Earth's atmosphere. The first attempt, in 1946, used a rocket assembled from captured German V2 parts. A crash-landing vaporized the rocket's payload, but the scientific potential of the project was still apparent. Since then X-ray astronomy has advanced along with space exploration. Observations made in 1978 and 1979 by an X-ray telescope aboard NASA's Einstein satellite have kept astronomers occupied for a decade.

Optical telescopes are also capable of an ever-increasing variety of observations. Light from a galaxy too distant to bring into focus can still be displayed on a video monitor, then sent through a spectrograph. Graphs of the spectral analysis can tell scientists the chemical composition, temperature, and behavior of stars millions of light years away. Soon astronomers expect to identify planets around some of the closer stars.

The esoteric forces and particles predicated by the theories of quantum physics require some unusual forms of measurement. At Homestake Gold Mine, a 100,000-gallon pool of perchlorethylene sits beneath a mile of South Dakota rock. The mixture of carbon and chlorine is familiar as cleaning fluid; however, it is used here to catch the elusive solar neutrino.

◀ A line of radio telescopes in Nagano Prefecture, Japan. The dishes gather radio waves transmitted from space and focus them on an antenna. One radio installation in rural New Mexico, the Very Large Array, consists of 27 dishes spread out over 500 miles.

Atomic reactions inside the Sun produce streams of neutrinos, particles that have no electrical charge and practically no mass. They pass unnoted and unchanged through atmosphere, flesh, and even lead, but they do interact with chlorine. So far, the percentage of altered chlorine molecules in the Homestake pool has been much lower than predicted by the equations of solar physics. Possibly the ratio of emissions to measurement is miscalculated, or perhaps astrophysicists will have to reassess some of their assumptions.

Some forces have yet to show themselves at all. The theory of general relativity includes a form of energy called gravitational radiation. The sudden movement of a mass, through a supernova explosion for example, creates a ripple in space that moves out from its source at the speed of light, causing a corresponding ripple in any object it encounters. Gravity waves are presumably ubiquitous, but they are very weak. They lose force as they diffuse outward, and an array of experimental measuring devices has so far failed to detect them. But scientists are persevering. They can now detect quivers of less than a quadrillionth of an inch in a supercooled bar of aluminum. When a celestial collision or explosion large enough to produce measurable gravity waves occurs, they hope to be ready.

▲ This manmade crater was produced by a simulated meteorite in a National Aeronautics and Space Administration laboratory. Researchers used high-speed cameras to record the impact of an aluminum bullet in a bed of sand. The raised margin of metamorphosed sand and metal shows in miniature the same process that excavated New Quebec and other craters.

probes have brought back photographs of craters on other moons and planets. Jupiter and Saturn are giant gaseous masses that do not show crater impacts on their miasmic surfaces, but their many moons do have impact craters. The Voyager probe revealed an 80-mile-wide crater on Mimas, one of Saturn's smaller moons, and a heavily cratered surface on icy Enceladus.

Craters were discovered on Mars and two of its moons by NASA's Mariner probes between 1965 and 1971. In 1972, radar observation revealed enormous craters on the cloud-shrouded surface of Venus. Tiny Mercury, which has almost no shielding atmosphere, is pocked with a multitude of overlapping craters.

METEORITE COLLISIONS

The theory of meteorite collisions was established partly by process of elimination. The other explanations, varying from extraterrestrial construction projects to volcanic eruptions, simply did not fit the geology of the craters. The lack of obvious meteorite craters on Earth was an initial drawback to the theory, but once scientists knew what they were looking for, they began to find craters on this planet.

As many as 100 million meteorites enter the Earth's atmosphere each day. The vast majority burn up and filter to the surface, if at all, as microscopic dust. Even those large enough to burn

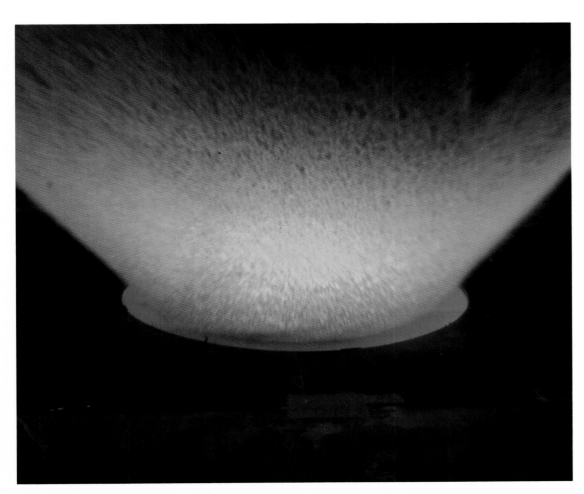

◄ The moment of impact. The energy from a collision at 18,000 miles per hour transforms the "meteorite" and the surrounding sand, sending up rays of molten material.

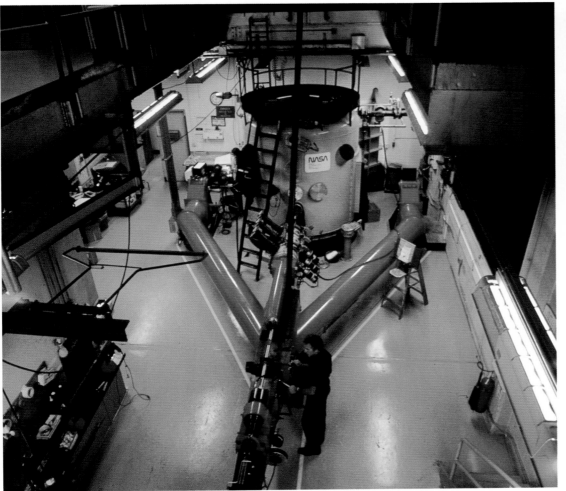

▲ Dr. Pete Schultz of Brown University alongside the sand bed used for a meteorite simulation

◄ Instead of the force of gravity, the NASA meteorite experiment uses a gun that can propel its projectile at five miles per second.

visibly as "shooting stars" have a diameter of less than 1/50th of an inch. About 500 meteorites per year reach Earth before being consumed, and of those, only about 10 are recovered. None within memory has been comparable to the colossus, an estimated 100 miles in diameter, that created the crater called Hellas on Mars. Yet Earth does still bear the scars of past collisions.

The first confirmed meteorite crater on Earth was identified in the Arizona desert in 1904 by David Moreau Barringer, a lawyer and mining engineer. Soil deposits in the crater date the impact at around 25,000 years ago, an estimate that concurs with references to the collision in a Hopi Indian legend. The meteorite responsible was about 100 feet in diameter. It has left behind a crater 4,000 feet wide and 600 feet deep. Barringer hoped to mine the meteorite for valuable minerals, but he had not taken into account the destructive power of the actual collision. Most of the meteorite was vaporized on impact. Its scientific value has outweighed any economic benefits.

Since Barringer's discovery, over a hundred meteorite craters have been identified on Earth. They represent a minuscule fraction of the number of terrestrial collisions, which are in turn only about 30 percent of the total. Most meteorites land in the oceans.

One of the largest and oldest craters lies on the west side of the Ungava Peninsula in northern Quebec. First identified from aerial photographs made by a prospector in 1950, the New Quebec Crater is two miles in diameter and 1,200 feet deep. All traces of the meteorite itself have been lost, but scientists estimate that it was the size of a football stadium and crashed into the earth at 50 times the speed of sound.

Sometimes the impact of a collision is so great that the ground actually recoils to form a mountain.

▶ A meteorite blasted out 300 million tons of rock to create the Barringer Meteorite Crater between Flagstaff and Winslow, Arizona.

THE EARTH'S INTERIOR

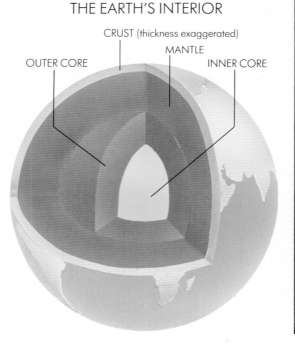

CRUST (thickness exaggerated)

MANTLE

OUTER CORE

INNER CORE

At Manicouagan Crater in Quebec, a central peak of 600 feet is all that remains some 200 million after the impact that formed it. The rings of mountains surrounding the lunar crater called the Mare Orientale were probably created in the same way.

The character of the rock around a crater gives another insight into the Earth's early history. Crater rock is often very similar to volcanic rock, since both are solidified from molten minerals. In the case of a volcano, the heat comes from beneath the earth. In the case of meteorite craters, it comes from the energy released by a massive impact. The collision also sends up a mushroom cloud of dust and grit, heated by the impact to hundreds or thousands of degrees.

Primeval Earth was covered with these craters. As the protoplanet grew bigger, its gravity exerted a

▲ Geologists travel to New Quebec Crater to read the story of the collision in the deformed and transmuted rock around its margins.

◀◀ New Quebec Crater on the Labrador-Ungava Peninsula. Some of the largest documented craters are in northern Canada, where a relatively stable geology and sparse vegetation have allowed them to remain visible.

▶ Meteorite craters have been identified on every continent.

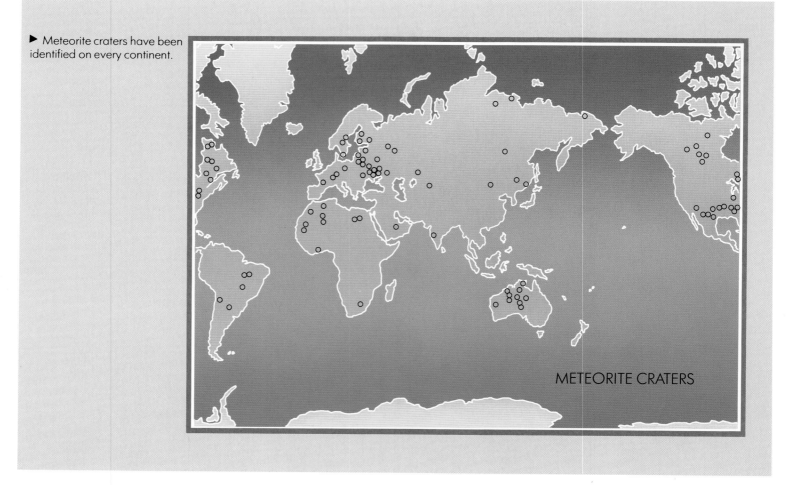

METEORITE CRATERS

▶▶ The central ring of Clearwater Lake Crater in northern Quebec is an indication of the violence of the meteorite landing. Rock heated to melting point by the collision splashed up into peaks around the point of impact. Images of lunar craters show similar kinds of formations.

stronger pull and meteorite collisions became more violent. The heat they created was a determining factor in the Earth's development. From a cold beginning, it developed a sea of molten rock over a cool interior. No life could have survived those early conditions, but without that season in hell, the Earth could not have evolved into the temperate haven of today.

Slowly the interior of the planet also began to warm, fueled by the decay of radioactive elements. Eventually the heat from within and impacts from without caused the Earth to separate into its chemical constituents. One was iron, a major component of meteorites originating in the inner solar system, which sank to become part of the molten metallic core. Another constituent was water, a component of some meteorites which was locked in until freed by heat. Still others were organic compounds, complex chemicals derived from carbon that are the building blocks of terrestrial life. Even amino acids have been recovered from meteorites.

All the planets in our solar system were created through the same processes of coalescence and collisions. However, the ensuing sequence of events on Earth made it unique. On Earth the iron and other dense elements began to sink below the surface, retaining their heat as they drew together at the planet's core. Radioactive minerals closer to the surface gave off more heat as they began to

degrade into stable elements. Water vapor, carbon dioxide, nitrogen, and other chemicals were released from their rocky prisons and rose to form thick clouds. The increasing force of gravity of the growing planet prevented these gases from escaping into space. This was the start of the Earth's present atmosphere.

AN ENVIRONMENT FOR LIFE

Gradually the surface began to cool and was covered by a crust of solid rock. Fewer planetesimals survived the friction of the atmosphere to reach Earth's surface. As the surface temperature fell, water vapor began to cool in the clouds above: this was the point at which the balance that led to life was struck. The vapor turned to liquid and torrents of rain began to fall on the still-hot surface, rising again as vapor and falling as rain once more. Billowing clouds reflected some of the Sun's radiant energy, lowering the surface temperature still more. The rains continued until the surface of the planet was flooded. Several hundred million years after its creation, Earth had oceans, land, and an atmosphere.

Countless meteors still enter our atmosphere daily, but we are generally only aware of them as the showers of "shooting stars" (actually microscopic specks). Only rarely do they survive to strike the ground. The only recorded human injury

LITTLE GREEN MEN

BEGINNING with the "Man in the Moon," and progressing to Isaac Asimov's stories and *Star Trek,* most of us grow up in the company of extraterrestrial beings. Whether we consider them friendly or frightening, the will to believe in extraterrestrial neighbors has adapted readily to the facts at hand. Christiaan Huygens, a brilliant Dutch astronomer and mathematician, discovered Saturn's largest moon, Titan, in 1655. Knowing that Saturn and its satellites take nearly 30 years to orbit the Sun, he observed that the Titanians' way of life "must be very different from ours, having such tedious winters."

Since Huygens's time the arguments for extraterrestrial life have both improved and weakened. We now know much more about the conditions that led to the evolution of life on Earth, and we know that at least some of those conditions were met elsewhere in our solar system. But increased knowledge also has eliminated some of our fondest fantasies.

Franz von Paula Gruithuisen, the astronomer who identified meteorites as the cause of lunar craters, also explained the cyclical appearance of a ghostly light on the dark side of Venus as the "general festival illumination in honor of the ascension of a new emperor to the planet." Later, he had second thoughts and decided that it was more likely to be firelight from slash-and-burn agriculture.

In fact, the light from fires on Earth has been photographed from space, though not from Venus. Neither the Moon nor Venus supports life — the Moon has no atmosphere, while Venus's thick clouds are poisonous and its surface temperature broiling. Of the other planets of the inner solar system, the favored candidate for biology is Mars.

Both H. G. Wells and Edgar Rice Burroughs popularized Martian life in a series of science fiction books. One of them, Wells's *War of the Worlds,* provoked panic when it was first broadcast on the radio in 1938. Although their books were fiction, both Wells and Burroughs took inspiration from a source at the pinnacle of American intellectual life. Percival Lowell, a Harvard graduate from a distinguished family, turned to astronomy after a career as a diplomat. He set up a telescope on a hill near Flagstaff, Arizona, now the site of Lowell Observatory.

Lowell's achievements were notable. He predicted the existence and position of the planet Pluto some 30 years before its discovery. But his main interest was Mars. Until his death in 1916, he studied, mapped, and wrote about the canals he saw there. Only a great civilization could have managed such a tremendous feat of engineering, and Lowell profoundly hoped that we could learn from their achievements.

The lesson in Lowell's work turned out to be very different. In 1965, when photographs came back from the Mariner 4 space flight, they confirmed what some astronomers even in Lowell's time had contended. Not only are there no civilizations on Mars, but there are no canals. Mars is a frigid, dust-scoured, crater-pocked desert. It has no flowing water. It is more hospitable than Venus or Mercury, but that is faint praise. Lowell's canals were figments of a predisposed imagination.

With the Mariner 4 photographs, most scientists' models for life on Mars shrunk to microscopic size. Biologists looked for clues in the valleys of Antarctica, the most Mars-like environment on Earth. Bacteria and algae have niches there, but these are pioneers from an immensely varied ecology developed in friendlier conditions. No one knows if they could have evolved independently. In fact, no one knows whether the preconditions of life on Earth — surface water, a nitrogen-rich atmosphere (the oxygen came later), organic chemicals, and moderate temperatures — were all necessary or simply sufficient. All life on Earth is based on carbon, an element that forms a variety of complex molecules that can power cellular metabolism. But is carbon necessary? Could silicon fill the same role? Is an ocean an essential incubator, or could smaller, more transient bodies suffice? We can hardly help generalizing from the only examples we know, but does that blind us to other pathways?

These issues came to the forefront in 1976, when Viking I and II space probes landed on Mars. While cameras scanned the landscape for the unlikely appearance of visible beings, experiments were carried out to check for microorganisms — or at least organic compounds — in Martian soil.

The difficulties of dealing with complex machinery from 40 million miles away made data hard to evaluate, but the final consensus was that the reactions on Mars were chemical rather than biological. None of the experiments located any organic compounds. These results did not completely rule out the possibility of life on Mars, but its scientific stock had fallen once again. Elsewhere in the solar system, the chances for life seem slim. Pluto, Neptune, and Uranus are all too cold.

Jupiter has possibilities, but in order to appreciate them we have to abandon our natural tendency to imagine life in a familiar image. Carl Sagan, whose freewheeling imagination sometimes drives more conservative scientists to distraction, has postulated (but not predicted) a Jovian bestiary. Jupiter's surface is a sea of liquid hydrogen at an unwelcoming -422 degrees Fahrenheit. However, parts of its 600 miles of atmosphere have Earth-like temperatures, water and a variety of other gases. Sagan and a colleague, E. E. Salpeter, proposed floating beings. Some might synthesize food from the atmosphere; others could prey on their neighbors. The project was more an exercise in breaking the conceptual bonds of Earth than a serious prediction of life, but it remains intriguing.

Notwithstanding its "tedious winters," some scientists also consider Titan a candidate for life. The Voyager I space probe in 1980 identified several organic compounds in its clouds. Back in the laboratory at Cornell, Sagan and B. N. Khare used the Voyager findings to brew up a version of the Titanic atmosphere. They ran an electric current through the nitrogen-methane mix, and waited. Soon a red film of complex organic molecules coated the walls of the test tube. In color and refracting properties, it matched the atmosphere on Titan itself. It was a chemical reaction, but it could be a first step toward biology.

Assuming there is no other life in our own solar system does not mean we are alone in the universe. We know that all the stars in all the galaxies are created of the same basic materials, and our Sun seems unremarkable as stars go. There is every reason to believe that the universe has other solar systems. Astronomers have begun to identify likely stars, and increasingly sophisticated measurements of their motion may soon reveal whether they are holding planets in their thrall.

Whether those planets have life will be harder to determine. The Pioneer 10 and 11 and Voyager 1 and 2 space probes are the first human hardware to leave the solar system, but light years will pass before they reach the next star. Since their arrival will be preceded by 60 years of radio and television programming, beaming inexorably into space, it may be that any civilizations they encounter will not be inclined to make their acquaintance. Just in case, however, the vehicles are equipped with messages.

The Pioneer space probes carry aluminum plaques engraved with a diagram of the solar system, their own route, a woman and a man, a binary numeral code expressing the wavelength of hydrogen, and other information that may be universal. The Voyager 1 and 2 spacecraft are each equipped with a phonograph record, *The Sounds of Earth* — ranging from cricket frogs to Chuck Berry's recording of *Johnny B. Goode* and a digital rendering of a photograph of a human birth. Whether or not the record comes to the attention of extraterrestrials, it is a valuable indication of how we see ourselves.

▲ Many early astronomers thought lunar craters like these were the creations of advanced civilizations.

▲ The energy of countless collisions and the heat of radioactive decay combined to turn the surface of the Earth into an ocean of magma. The gases released by the melting rock helped to form the planet's secondary atmosphere.

from a meteorite took place in Sylacauga, Alabama, in 1954, when a three-pound meteorite crashed through the roof of a home and struck Mrs. E. Hulitt Hodges, bruising her left thigh and interrupting her nap.

A more consequential event occurred in Siberia in 1908, when a fireball exploded some five miles above the ground, destroying 1,200 square miles of timber and singeing the clothing of people 60 miles away. Photographs of the forest after the blast resemble the slopes of Mount St. Helens after its eruption in 1980. No impact crater has been found, but E. Sobotich, a Soviet scientist, has found a sprinkling of tiny diamonds at the site.

Most scientists think the fireball was part of a comet that exploded when it reached the atmosphere and melted before it reached the ground. A comet is often described as a dirty snowball, with frozen methane and ammonia as well as water.

In 1969, a meteorite fell in northern Mexico, lighting the sky on its approach and fragmenting into jagged chunks on impact. The Allende meteorite, named after the village where it landed, was a link to the earliest days of the solar system. Scientists at Case Western Reserve University found evidence of radiation damage in the rock that probably originated in the protosun more than

4.5 billion years ago. From that time until entering the Earth's gravitational field, the planetesimal had orbited untouched through the solar system.

These dramatic visitations are rare. However, some theories hold that the arrival of meteorites comes in cycles. If so, our distant descendants, if any, may be in for a rocky time some 13 million years from now. According to one hotly debated hypothesis, the Earth enters an impact period about every 26 million years. Some paleontologists have linked major periods of extinctions indicated in the fossil record to such a cycle. Others doubt the existence of a cycle but do think that the mass extinctions that included the end of the dinosaurs

about 65 million years ago might be attributed to the impact of a massive meteorite or a swarm of smaller collisions.

One scenario goes like this. A giant meteorite, perhaps as much as six miles in diameter, is drawn into the Earth's orbit and plunges toward the ocean. Barely slowed by thousands of feet of water, it hits the bottom within a fraction of a second and explodes, generating temperatures of 32,000 degrees Fahrenheit, hotter than the surface of the Sun. A huge mushroom cloud of steam and pulverized rock rises up through the "hole" made by the meteorite's passage through the atmosphere and spreads quickly around the globe. A tsunami

▲ Clouds rise from the unstable surface of the young Earth. When the atmosphere cooled enough to condense water vapor into rain, the planet entered a new phase of development.

▲ A fireball over the small town of Pueblito de Allende in Chihuahua, Mexico, heralded the largest shower of stony meteorites on record on 8 February 1969. The explosion scattered an estimated four tons of fragments.

wave 200 feet high batters the coastlines and churns up tons of sediment. On land, the heat of the collision sets off firestorms, using up oxygen, generating deadly carbon monoxide, and sending still more smoke and dust into the atmosphere.

This series of events resembles in many ways the projections of the ecological catastrophe that could follow nuclear warfare, although the energy released would be several times the combined

nuclear forces of the United States and the Soviet Union. Many plants and animals would perish in the firestorms, or survive only to find their habitats destroyed. The dust cloud would block the passage of sunlight, stopping photosynthesis for as long as three months. Without sunlight, the ubiquitous phytoplankton in the oceans would die, followed by the marine invertebrates that depend on them for food. Many land-based plants would also

◄ This meteorite at the Mexico City Museum weighs 14 tons and is almost pure iron. Its composition indicates that it was once at the dense, metallic core of a much larger planetesimal.

▼ The people of Pueblito de Allende were roused from their beds by a light "as bright as day" and a violent gust of wind. They recovered meteorite fragments weighing up to 50 pounds and reported that even after their fiery journey the rocks still held the chill of space.

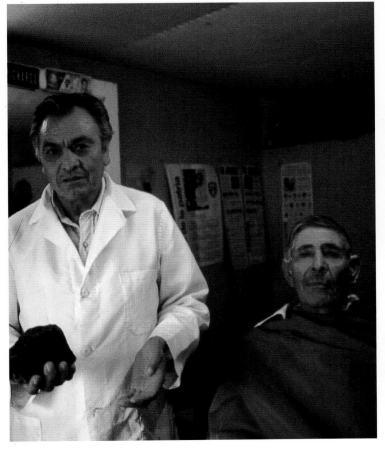

▼ ▶ Although most of its species are extinct, the world of the Cretaceous can be reconstructed in part from fossil plants and animal bones. Some plants have been preserved so faithfully that paleontologists can examine their cell structure.

▶ Their abundant fossil remains make the coiled ammonites and squidlike belemnites among the best-known casualties along the K-T boundary. After flourishing for more than 200 million years, they vanished along with hundreds of other marine species.

▶▶ At the top of the food chain, *Tyrannosaurus rex* and other dinosaurs could have been rendered extinct either directly by a change in climate, or through starvation with the death of countless smaller organisms.

succumb to a combination of dark and cold. When the dust cleared, sunlight would return to an impoverished world.

K-T BOUNDARY

In the fossil record, the end of the dinosaurs, along with more than half the other known animal species on Earth, marks the boundary between the Cretaceous and Tertiary periods. Paleontologists have shortened the term to "K-T boundary." In the Tertiary period the evolutionary torch passed to mammals, whose species increased dramatically in size and diversity. If the K-T boundary records a meteorite collision and is not just an arbitrary label imposed on an evolutionary continuum, there should be some evidence in the geologic record. Given the Earth's tumultuous physical history, researchers cannot count on finding the actual impact crater; however, an event big enough to render hundreds of species extinct would leave other traces.

In 1979, a University of California/Berkeley geologist named Walter Alvarez took samples of

the Gubbio layer, a section of sedimentary rock formed in the ocean and more recently elevated into the Appenine Mountains north of Perugia, Italy. A thin layer of clay separates limestone from the Cretaceous and Tertiary periods. Compared with the Cretaceous limestone, the Tertiary sample

NEMESIS CONTROVERSY

THE theory that some kind of collision with an extraterrestrial object precipitated the demise of the dinosaurs is still being debated. The controversy continues to attract an unusual degree of public attention. In the process, it also highlights the practice of science in ways both inspiring and unsettling. It is the job of science to develop new theories, then seek out evidence for and against. Many scientists have been invigorated by the extinction debate and the disciplines it has brought together. Others have been bruised by a controversy that has at times become highly personal.

Scientists, like everyone else, tend to cling to old verities and to feel territorial about their specialties. Some of the opposition to the meteorite extinction theory clearly springs from these impulses, but the theory is also open to legitimate criticism. Fossil records are incomplete and dating techniques are less than precise, so it is difficult to prove a relationship between a collision and an extinction. Dinosaur species, never numerous, may have been on their way out already due to longterm changes in climate. Conversely, the discovery of dinosaur fossils in northern Alaska has led to speculation that at least some species could have adapted themselves, as many surviving mammals did, to a season of polar darkness.

Still, it is hard not to conclude that some major event took place at the K-T boundary. Dinosaurs are the celebrities of the Cretaceous, but hundreds of other species died as well. The geological oddities in the boundary layer also require some kind of explanation.

One alternative to a meteorite is a period of intense volcanic activity, which could produce many of the same effects. Carbon dioxide released by the eruptions might have contributed to a warming trend that eventually became lethal. The iridium anomalies could be the result of biological processes much as calcium carbonate being concentrated by marine organisms.

Rather than settling the issue either way, each new piece of information seems to intensify the disagreements. Dr. Robert Jastrow, a Dartmouth astrophysicist, is among those who emphatically reject the meteorite concept, telling the *New York Times* that "a catastrophe of extraterrestrial origin had no discernible impact on the history of life as measured over a period of millions of years."

Luis Alvarez, one of the originators of the theory, responded that Jastrow "is not a very good scientist." Alvarez also called Dr. Dewey McLean, a leading proponent of a volcanic disaster, "a weak sister."

In 1983 a group of scientists raised the stakes on the meteorite theory by proposing first that major extinctions follow a regular cycle, and then that the cycle is linked to some kind of galactic force. The idea is that Earth periodically comes into the range of heavy bombardment by meteorites or comets. Among the proposed agents of this barrage are the orbital influence of a tenth planet located somewhere beyond Pluto; the vertical oscillation of the solar system as it moves through the Milky Way; and the effect of a companion star to the Sun.

The proposed companion star, named Nemesis, has garnered the most popular attention. In outline the hypothesis is this:

Nemesis has a highly elliptical orbit around the much larger Sun. As it passes through a cloud of dust and debris called the

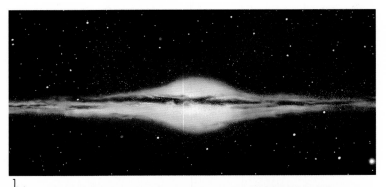

1.

Oort Cloud in the outskirts of the solar system, its gravitational pull dislodges planetesimals from their own orbits. Most of them head harmlessly into space, but some collide with Earth.

Since Nemesis' orbit follows a 26-million-year cycle, these periods of impact also occur about every 26 million years. It was one of these collisions that destroyed the dinosaurs and thousands of other species 65 million years ago. Another, 250 million years ago, killed 95 percent of the species then living.

There is no hard evidence in favor of Nemesis. It is simply one theoretically possible explanation for periodic extinction events which are themselves unproven. But such hypotheses are immensely useful for science, providing a launching point for experimentation and debate. Astrophysicists have gone to work on Nemesis' orbit, testing equations to assess its stability and plausibility. Astronomers are looking through star catalogs for likely candidates.

Paleontologists and geochemists are analyzing fossils and rock samples in the light of a new set of assumptions. Opponents to the theory are gathering data to support their own alternative explanations. Whether or not Nemesis is ever found, other discoveries will no doubt result from the search.

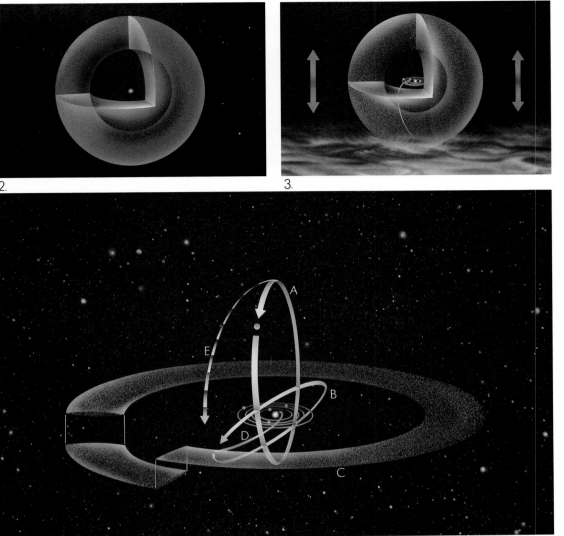

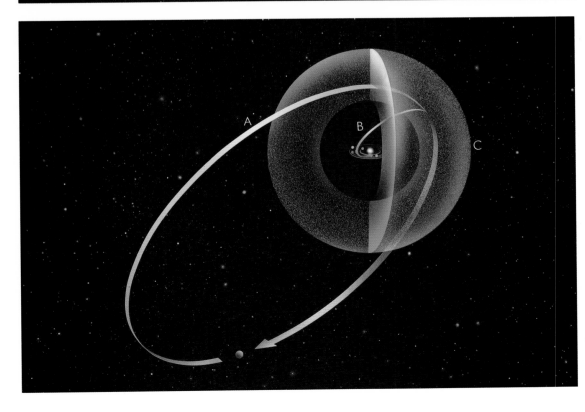

1. The Milky Way galaxy is a vast disk, some 100,000 light years across, trailing spiral arms of stars.

2. Our solar system crosses the horizontal plane of the disk about every 31 million to 33 million years.

3. As the galaxy is densest along the plane, the Earth might be subject to more meteorite bombardment and more radiation as it passes the midpoint. This theory has the advantage of dealing with an already accepted premise; however, the estimated time for the next pass through the galactic plane does not coincide with the next predicted period of bombardment, a discrepancy that must be reconciled before the theory finds acceptance.

◄ Planet X theory

A. The orbit of Planet X.

B. The orbit of Planet X is caught in Oort's cloud.

C. Oort's cloud surrounding the solar system in a donut shape.

D. A comet is sent into the center of the solar system off from the orbit.

E. The orbit of Planet X changes at a cycle of 28 million years.

◄ Planet X would be expected to have an orbit far out of plane with the known planets of the solar system. About every 26 million years it would pass through a disk of comets located beyond Neptune. Although centuries of observation have failed to locate such a body, the existence of Planet X could explain some demonstrated oddities in orbits of the outer planets.

◄ Nemesis theory.

A. The orbit of companion star.

B. A comet is sent into the center of the solar system off from the orbit.

C. Oort's cloud.

◄ In order for Nemesis to play its postulated role in periodic extinctions, it would need to have a highly eccentric orbit and about one-tenth the mass of the Sun.

▶ Dr. Frank Asaro monitors the chemical analysis of K-T boundary sediments at the Lawrence Berkeley Laboratory. Elevated levels of iridium were an unexpected finding during an attempt to establish deposition rates of various types of sedimentary rock. "We were very surprised," said Asaro. "You couldn't explain that much iridium coming from the surface of the Earth . . ., so we said it had to come from outer space."

shows a drastic drop in the amount and diversity of marine plankton. Something happened to the environment in the period between those two layers. Using a sophisticated system of chemical analyses overseen by Alvarez's father, Luis Alvarez (a Nobel laureate in chemistry), the clay in between provided one clue. Its proportion of iridium was 160 times higher than in the limestone above and below.

METEORITE EXTINCTION THEORY

Iridium is a very dense and very rare element found in meteorites. Measured in parts per billion, it comprises a minuscule portion of the Earth's makeup and, due to its density, is mostly in the Earth's mantle, not in its crust. One source for an unusually high proportion of iridium could be debris from a large meteorite.

Opponents of the meteorite extinction theory point out that volcanos are another source of iridium. They can bring up relatively iridium-rich material from the Earth's mantle and can also send clouds of debris into the atmosphere. Evidence of major volcanic activity around the time of the extinctions has been found in India.

Whatever its cause, another record of the K-T boundary is found in chalk cliffs of Stevn's Klint, on the Danish coast south of Copenhagen. There is a narrow blackish layer between the white chalk of the Cretaceous and the white limestone of the Tertiary above. The dark layer is called fish clay after the fossil fish bones that were found in it, but its hallmark is that it is otherwise devoid of fossil life. Clay, unlike chalk and limestone, is

nonorganic. The Stevn's Klint clay also has elevated levels of iridium. Above the fish clay, the fossil record shows none of the ammonites, belemnites and other Cretaceous creatures that are preserved in the chalk below the boundary.

More than 50 sites around the world have shown elevated iridium levels. Since 1980, similar layers

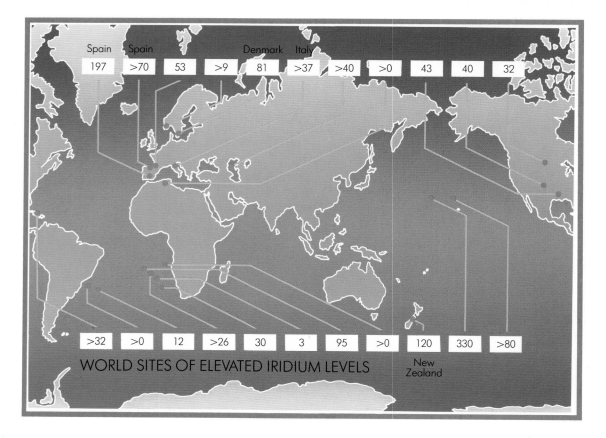

	Spain	Spain			Denmark	Italy					
	197	>70	53	>9	81	>37	>40	>0	43	40	32

	>32	>0	12	>26	30	3	95	>0	120	330	>80

WORLD SITES OF ELEVATED IRIDIUM LEVELS   New Zealand

◀▲ The cliffs at Stevn's Klint, Denmark, display clearly the layer of "fish clay" separating the Cretaceous and Tertiary periods. Dr. Finn Surlyk of the Geological Survey of Greenland, at right, has studied the layer for clues to the K-T extinction.

THE UNIVERSE — OPEN OR CLOSED?

THE birth of the universe, once a matter of myth and metaphysics, is moving into the realm of the demonstrable. Since 1965, when Arno Penzias and Robert Wilson recorded the remnants of light from the Big Bang, scientists have found more ways to match theory with observation. Alan Guth, a theoretical physicist at the Massachusetts Institute of Technology, has come up with a widely studied mathematical model of the cosmos's first milliseconds, when an unimaginably dense collection of matter exploded with a force that still powers all the stars and planets. Timelines like Guth's are highly speculative, but scientists are getting better at gathering evidence about them. Something happened some 15 billion years ago, and it left records which we are beginning to decode.

Even more speculation surrounds the question of how the universe will end. We have learned enough about the birth and death of stars to project the Earth's demise billions of years hence. If some other human or cosmic disaster does not overtake it first, the Earth will be burnt to cinders as the Sun runs out of its hydrogen fuel and expands to enter a new phase as a red giant star. But the Sun is only one star among trillions. What about the fate of the universe as a whole?

The universe is expanding. Every galaxy is moving away from every other galaxy. Either it will expand forever, a scenario scientists call open, or the pull of gravity will eventually exceed the outward force of expansion and the direction will reverse.

In fact, the expansion, which can be measured by the shift in light waves as the source of light moves, is already slowing. Physicists are trying to determine what the distant future will hold. According to Einstein's theory of general relativity, the motion of the universe depends on how much matter it contains and the concomitant force of gravity. If there is enough matter, gravity will eventually overpower expansion. Like a videocassette on rewind, the whole process will go into reverse. If the gravitational pull is insufficient, the expansion of the universe will continue forever. Estimates of the mass of the universe fall on the borderline between these two scenarios, and each has its adherents.

The universe is still well supplied with the interstellar gases that are the formation grounds of stars. But in about 100 billion years, the galaxies will be dying as star after star uses up its fuel. If the universe is open, the cosmic corpses will go on receding forever. In their death throes, the densely packed stars in the galactic cores will coalesce into black holes with masses hundreds or thousands of millions times greater than the Sun. Black holes are the stuff of science fiction, but they are also a theoretical consequence of general relativity. They occur when compression creates a gravity force so extreme that not even light can escape. Anything that comes too close to a black hole's force field, its "event horizon," will vanish forever.

In an expanding universe the black holes themselves will grow larger as they consume more and more, absorbing all the matter presently seen in the galaxies and then even the dispersed particles in interstellar space. After hundreds of billions of years, the black holes themselves will evaporate. That process will produce particles and antiparticles whose reaction will create a new presence of radiation, but it too will become ever weaker and more thinly dispersed. The ultimate end of worlds will come not with a bang or even a whimper, but with the endless expansion of empty space.

If the universe is closed, expansion must eventually stop. Once it does, gravity will begin to pull everything together again. The redshifted lightwaves that signal movement away from the viewer will become blueshifts. As the galaxies come closer, they will begin to collide and merge. The night sky will become solid with stars, and more stars will be formed by the newly compressed gases. Radiation, some of it from starlight trapped by increased gravity, will increase in intensity, transforming visible light into ultraviolet radiation, and then into X-rays and gamma rays. The remnant radiation discovered by Penzias and Wilson will start to heat up again.

Eventually heat and radiation will attack the foundations of matter, destroying even atomic nuclei. As the universe is squeezed into an ever decreasing volume, the matter and radiation that have escaped the black holes will be squeezed out of existence. The entire universe will be compressed into a singularity, an incomprehensible condition of infinite density and no volume. Physicists call this process "The Big Squeeze" or "The Big Crunch."

Neither model is entirely satisfactory, either scientifically or philosophically. There is something profoundly unsettling about imagining the universe first created out of nothing, and later vanishing altogether. Some scientists simply leave the ultimate beginning and end to God. Others have developed theories that allow for eternal expansion and contraction.

In 1922, when the underpinnings of particle physics were still new, a Soviet scientist, Alexander Friedmann, proposed an oscillating universe that went through cycles of expansion and contraction with recurrent big bangs. His calculations were expanded in the 1930s by Richard C. Tolman in the United States. One problem with that theory is the presumable accumulation of radiation through a series of universes, which would make each cycle bigger than the one before and impose a limit on how many cycles were possible. This brings us back once again to the question of a beginning.

To circumvent that problem, a British physicist named Thomas Gold proposes that when the universe reaches its point of maximum expansion, time begins to flow backward. Starlight returns to stars, black holes relax their grip on the matter trapped within them. The Earth returns from the ashes as a land where rivers run uphill and animals progress from death to conception. All this seems manifestly impossible, but there is a catch. Since our mental processes also would reverse, we would experience time as we do now. Perhaps this book is on its way back to the printer, and hence to the forest. Wild though it sounds — and not many scientists are persuaded — Gold's theory is capable of experimental proof and of modifications that may make it more plausible.

One variation, the "superspace" theory proposed by University of Texas physicist John Archibald Wheeler, holds that the laws of physics we know are simply a function of the present cycle. With each new cycle the eternal verities may change. Gravitation may function differently, and constants such as the speed of light and the charge on an electron may vary. Perhaps the physical realities we know may simply break down under the extreme conditions of singularity, and our assumptions about matter will be irrelevant. Yet another theory is that black holes are the compressed seeds of new universes, each of which develops in endless isolation from the others.

▲ A dark nebula, where gases are compressed into an intensely radiant mass.

have been identified in several spots including Tokachi Plain on the Japanese island of Hokkaido and Caravaca in Spain.

In Spain, in addition to the black clay, there is a narrow, reddish band that may tell something about the location of a collision. The red vein contains materials formed when rocks which were once on the ocean floor were melted by an extremely powerful impact and explosion. In 1988, scientists from the University of Washington announced a related finding. Along the banks of the Brazos River in Texas, they found yard-thick beds of clay chunks and sandstones that had been deposited at the time of the K-T boundary. According to Joanne Bourgeois of the UW team, they were left by a huge tsunami wave, which could have been caused by a meteorite landing.

Soot has also been identified in clay layers in Spain, as well as in Denmark and New Zealand. Though not found in definitive quantities, it may be

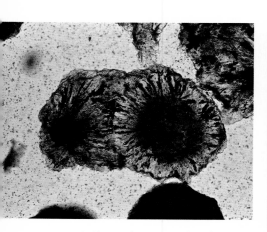

▲ These glassy particles, shown under magnification, may be microtektites created by a meteorite landing. They have been found not only at the K-T boundary but in other layers that coincide with major extinctions. Although larger tektites are generally agreed to be caused by meteorites, the derivation of microtektites is less clear and therefore has not been conclusive in establishing periodic extinction.

▶ *Right and far right:* The hills of Caravaca, in southern Spain, were once beneath the sea. They show a layer of fish clay similar to that at Stevn's Klint, as well as a narrow reddish band that may include remains from a meteorite explosion. It includes tiny droplets of a type that is formed by the condensation of vaporized minerals. However, some geologists have reported finding similiar droplets in sediments well above and below the K-T boundary, casting doubt on their connection with a particular explosion.

A Meteorite Strikes

▶ 1. A meteorite enters the atmosphere, beginning to burn about 40 miles above Earth's surface and pushing air in shock waves in front of its path.
2. Seconds later, it crashes.
3. The energy of its motion is converted to heat, causing an explosion.
4. Rocky fragments, pulverized debris, and superheated gases are flung skyward.
5. As the heavier fragments fall back to earth, dust and gases rise in a mushroom cloud.
6. Debris continues to rise and spread across the upper atmosphere as the new crater cools.

1.

2.

3.

4.

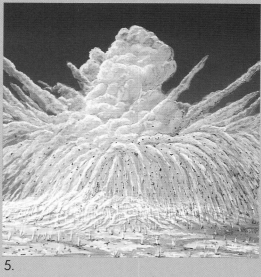

5.

6.

the record of firestorms caused by the meteorite. Additional evidence comes from the presence of shocked quartz, a rock form associated with meteorite impact, in the boundary layer.

If a meteorite collision did cause the extinctions at the end of the Cretaceous period — and despite accumulating evidence this is by no means proven — was this an isolated event or part of a pattern? Is it possible to predict what might happen next?

At an astronomical observatory on top of Mount Palomar in California, Eugene and Carolyn Shoemaker have studied asteroids for 25 years. Using an astronomical telescope fitted with a highly sensitive camera lens, they have spotted two dozen bodies that could come close to Earth's orbit. If any one of these were to collide with Earth as a meteorite, the consequences would be disastrous.

"There very definitely is a chance that a ten-kilometer asteroid will collide with the Earth in the future," says Eugene Shoemaker. "They've hit about once every 50 million years, on a long-term average. . . . We actually know of two Earth-crossing asteroids right now . . . that are about that size, each of which could hit the Earth."

One of the asteroids being tracked by the Shoemakers is Eros, which has about a 20 percent chance of hitting the Earth in the next several hundred million years. At about 12 miles in diameter, Eros is twice the size of the putative K-T meteorite.

Astronomers can deal only in long-term probabilities and not in short-range predictions. Rather than seeing each shooting star as a potential destroyer, we might just as well marvel at how its predecessors formed our planet and brought us the stuff of life.

▲ Carolyn and Eugene Shoemaker use an 18-inch telescope at the Palomar Observatory to track comets and asteroids. Most summers they travel to Australia to study craters.

◄ The distinctive topography of Iceland is a creation of the Earth's internal heat. These pockmarks, called pseudocraters, occur when lava flows over waterlogged soil. The water is converted to steam that forces its way through the molten covering.

THE HEAT WITHIN

BENEATH the Earth we know lies an inferno. Sometimes, its skin cracks, and we come face to face with the heat of its heart. One of the sites of these revelations is the Ring of Fire, a line of volcanos along the unstable Pacific coast from Tierra del Fuego to Alaska, and from the Kamchatka Peninsula south again to New Zealand.

Mount St. Helens in Washington state is one of these peaks. One of the youngest mountains in the relatively young Cascade Range, it probably developed its famous symmetrical cone during the past millennium. And although scores of postcards and snapshots have celebrated its serenity, it is not dead.

The Klickitat Indians who lived nearby called it Tah-one-lat-clah, Fire Mountain, and prudently stayed off its higher slopes. Despite a fiery eruption in 1842, their successors were less cautious. Communities took root on its flanks, and vacationers flocked to its hiking trails and splashed in the chilly waters of Spirit Lake.

Then in March 1980 the mountain stirred. Geologists noted first an increasing number of earthquakes, then some minor eruptions. Adventurers hiked up to see the smoke and feel the mountain jiggle. Some of them paid with their lives. On the morning of 18 May 1980, at 8:32 a.m., the summit exploded with a force 500 times that of the atomic bomb that fell on Hiroshima, losing nearly 3,200 feet of its northern rim.

Whole hillsides of timber were flattened. Spirit Lake sloshed, boiling, halfway up the opposite slope. A debris cloud rose into the stratosphere, blanketing thousands of square miles with choking ash. Inside the volcano the temperature topped 1,500 degrees Fahrenheit. Local residents, who

take a proprietary pride in their mountain landscape, were reminded that the peaks are not mascots but masters.

The heat released by the eruption of Mount St. Helens comes from the Earth's mantle, the product of forces that are still shaping our planet. For Earth is still an evolving world, where continents move and volcanos give birth to new lands. Many of its secrets are buried beneath the surface.

The bottom of Western Deep Levels Mine near Johannesburg, South Africa, is the deepest point

▲ Huge refrigeration units provide the air conditioning and cooling spray that make work possible at the bottom of Western Deep Levels mine.

◄◄ The ravaged summit of Mount St. Helens testifies to the violence of its eruption in 1980. In the background is Mount Adams, which bears the marks of many past eruptions.

▲ Steam issues from a fissure line atop the Reykjanes Ridge in Iceland. Underground water is heated by magma rising from the Earth's mantle.

that humans have reached beneath the land's surface. Gold miners take a half-hour elevator ride to reach their jobs 11,736 feet below and giant refrigeration units keep air temperatures barely tolerable as they drill into rocks hot enough to burn the naked hand. Mines like Western Deep Levels are impressive engineering feats, but they barely nick the Earth's crust, which is 40 miles thick in some spots. The average temperature of rocks on the Earth's surface is about 55 degrees Fahrenheit. Western Deep Levels miners cope with

temperatures 70 degrees higher. At 250 miles below the surface, in the Earth's mantle, geologists estimate temperatures of over 2,500 degrees. The core is even hotter — perhaps as much as 11,000 degrees Fahrenheit.

During the planet's formation, countless planetesimals collided with it and caused its surface to heat over a relatively cool center.

While the Earth was still young, radioactive decay heated the interior of the planet until it became molten. The heaviest elements in the molten rock sank, creating a core of iron and nickel. Less dense but more radioactive material concentrated in the mantle, where it continues to produce heat, while basalt and other relatively light rocks cooled to form an insulating crust.

REMAKING OF EARTH

Radioactive heat still powers some of the most important workings of the planet. In some spots it remakes the land right before our eyes. One such place is Iceland. Created and sculpted by volcanic action, the North Atlantic island still averages one major eruption every five years. Beneath the glaciers that give Iceland its name is an oceanic ridge, a spreading zone where the sea floor is being pushed apart, inch by inch, by lava pushed up from the mantle.

The process is slow, taking nearly 5,000 years to build up three feet of volcanic rock, but over the

ICELAND

Thingvellir
Reykjavik

Lakagigar Crater

Surtsey Island
Heimaey Island

past 65 million years eruptions and erosion have combined to make an extraordinary landscape. Beneath the glaciers, hot springs create underground rivers and pools where bathers can lounge in comfort in a cave of ice. Above the glaciers, a line of volcanic fissures is plainly visible from the air. The most famous is Thingvellir, a dormant section of fissure that has cooled and shrunk to form a narrow, straight-sided valley. A giant emerald-green trench, Thingvellir looks more like an engineering project than a natural formation. Elsewhere on the island, glaciers have scraped away the surface coverings of lava flows, revealing architectural ramparts of hexagonal basalt columns formed in the interior of the flow as it cooled.

Icelanders have learned to coexist with their unpredictable topography, moving elsewhere when lava comes too close and returning as soon as the landscape stabilizes. For over 800 years Thingvellir was the meeting place of their legislative assembly, the Althing, and the site of the drowning pool traditionally used to eliminate lawbreakers. More recently, Icelanders have tapped geothermal energy for a host of practical uses from industry to home water heating. As forecasts and evacuation procedures grow more sophisticated, the likelihood of the kind of volcanic disaster that killed 10,000 people and half the island's livestock in 1783 diminishes. But danger remains.

In 1973, a violent eruption reshaped Heimaey, one of the Vestmann Islands off the southern coast of Iceland. Most of the 5,000 inhabitants were evacuated and more than 350 buildings were burned or buried when a 735-foot volcano grew in the fields of a coastal dairy farm. Even worse, a

▲ Where magma approaches the surface, heated pools form beneath a covering of glacial ice. Eruptions on glacier-covered volcanos have produced devastating floods as the ice suddenly melts.

lava flow threatened to close off the only entrance to the harbor, home to the country's most profitable fishing fleet.

Their boots smoking as they walked across a thin crust of new rock, an army of volunteers pumped six million tons of sea water onto the encroaching lava, trying to slow its advance. It seemed an unequal contest, but the Islanders managed to outlast the eruption. When the new volcano subsided five months later, the harbor entrance had narrowed to 150 yards. A thick, gritty layer of volcanic ash covered every surviving house and street, but most of the Islanders returned. They used the volcanic leavings to make another runway for their airport and built a lava-heated municipal swimming pool.

Iceland's major volcanos lie along a line extending from the northeast to the southwest. Aerial surveys show that the island's fissures, long cracks in the Earth's surface, are oriented along the same axis. Both continue off the land into the

▲ The fissure line at Krafla, at the northern end of the Reykjanes Ridge, was dormant for 250 years after major eruptions in the eighteenth century. In 1975 the fissure line again began spewing fire, and the crack widened 15 feet in less than a decade.

◄ These basalt pillars are a sign of a volcanic landscape. Under certain conditions, lava cools into hexagonal columns.

▲ A change in topography marks the collision point between India and Asia 50 million years ago.

▶ Icelanders have learned to live with uncertainty. Homes and roads are built alongside a dormant fissure.

North Atlantic. The waters south of the main island are dotted with small islands like Surtsey. Named after Surter, a tenth-century Norse giant who was said to control fire, Surtsey was created in 1963 in an underwater eruption. The island was as barren as the newborn Earth at the outset, but colonists arrived almost immediately. By 1968, botanists had recorded over 100 species of photosynthetic bacteria, and Surtsey is now home to nesting birds and flowering plants.

UNDERWATER VOLCANOS

The extent of underwater volcanic activity was difficult to gauge until recently. In 1973 a survey jointly conducted by France and the United States investigated the floor of the Atlantic Ocean off the Azores, to the west of Portugal, islands which were themselves formed by undersea eruptions. The vessels — *Cyana, Archimède,* and *Alvin* — descended 10,000 feet below the surface. In the

PLATE TECTONICS

Over the centuries science has confounded our intuitive belief in Earth as the center of the cosmos and man as the measure of all things. We have come to accept that the Earth is round; that it spins through space in consort with other planets; that the life upon it is not immutable but is evolving in unknown directions.

In 1912, a German meteorologist named Alfred Wegener asked us to consider that the very ground beneath our feet was in motion, sliding across the ocean floor and constantly recreating our geography. An American geologist, F. B. Taylor, came to the same conclusion two years earlier, but it was Wegener's book *Die Entstehung der Kontinente und Ozeane,* that spelled out the theory and suggested tests that might confirm it.

The hypothesis, however bizarre it sounded, was not new. As early as 1756 another German, the Reverend Theodor Lilienthal, noted that "the facing coasts of many countries, though separated by the seas, have a congruent shape, so that they would almost fit one another if they stood side by side." The striking correlation between southern Africa and the Brazilian coast, in particular, was apparent to anyone who had a map. And as more travelers visited both spots, they found other similarities in geology, flora, and fauna. A century later the theory of evolution brought new sophistication to the study of fossils, and still more similarities were found.

In the late nineteenth century, Edward Suess, an Austrian geologist, posited an ancient continent he called Gondwanaland, which encompassed all the continents of the southern hemisphere, plus the Indian peninsula. His theory had a necessary corollary, a giant northern continent given the name Laurasia. Evidence mounted in support of Suess's theory, but the transition from Gondwanaland to the contemporary world was still entirely unclear.

Wegener suggested an answer. He said that 300 million years ago, all the Earth's land was joined in one supercontinent to which he gave the name Pangaea. Over the millennia, Pangaea's surface cracked, sank, and eventually separated into several segments. These fragments, which Wegener called continental plates, began drifting across the ocean floor, a process which is continuing even now. Wegener cast the net of this monumental supposition over some of science's thorniest puzzles:

MOUNTAIN RANGES: According to Wegener, mountain ranges were formed by compression and folding, as continents approached one another and then collided. Until then most scientists thought non-volcanic mountains were the byproduct of a cooling Earth that cracked and shriveled into ridges and peaks, rather like the transformation from grape to raisin. But the Earth would have to be cooling at a tremendous rate to produce even a modest peak, and other geological evidence did not support a rapid drop in temperature.

ANIMALS: Some of the earliest support for Suess, and later for Wegener, came from the worldwide distribution of species. Why, for example, would Lumbricidae, a family of snail, be found in Japan, Asia, Europe and the east — but not the west — coast of North America? Could a creature cross the oceans but not make its way across the continent? In Wegener's map of Pangaea, the snail's current homes are dovetailed into one area. Fossil records show many other striking congruences.

MINERALS: The geology of South Africa and Brazil not only looks similar, but contains the same elements in the same formations. Likewise, the Caledonian System of Northern Europe corresponds to the Appalachians of Canada and the United States. And it is also persuasive that these matches are not found in mountains formed after the Cretaceous period, when Wegener believed that Pangaea split apart.

CLIMATE: Evidence of former glaciers has been found near the Equator, and coal deposits show that Antarctica once supported abundant vegetation. Wegener's construction of Pangaea places these continents in the right place during the right eras to have accumulated these artifacts.

For all its elegance, however, Wegener's theory had one major weakness. To set the continents adrift and slide them across the sea floor to the top of the world requires a mighty force. The ancients could believe that the Earth moved when Atlas shrugged. Twentieth-century scientists needed something more. Wegener provided an explanation, based in part on the tidal attraction of the Sun and the Moon, but it seemed unlikely that they were sufficient for the task. Calculations soon showed that friction would stop the continents from sliding across the ocean floor. After a period of great excitement and controversy, continental drift fell into disrepute in scientific circles. Wegener died on an expedition to Greenland in 1930, leaving the theory without its most persuasive proponent. By the 1950s its stock had fallen so low that the *Encyclopedia Britannica* published in 1957 dismissed Wegener's construction of Pangaea as "purely fanciful."

But the tide was about to turn. Research in a number of fields produced evidence supporting Wegener's contentions that the continents had moved. New understanding of radioactivity as a source of energy indicated that the Earth was not cooling, but might even be heating up. This further undermined competing theories of mountain range formation, and also suggested a force strong enough to split Pangaea apart. Even more dramatic was the evidence provided by measurements of magnetism. Lava traps a record of the Earth's magnetic field as it cools, and those records suggest that continents had once been in different locations, and that they had moved during the period postulated by Wegener.

In the 1970s, the first sophisticated mappings of the ocean floor showed the outlines of the plates on which the continents rest. The big flaw in Wegener's hypothesis was overcome by seismic measurements that revealed a soft, plastic layer of rock about 60 miles below the surface of the Earth. The outer crust slides across this deep, slippery layer.

More evidence emerges daily, on lava fields, in research submarines, in computer printouts. The conclusions reached by Wegener in a tremendous leap of intuition are being corroborated in ways that not even he could have guessed.

Still, some geologists find the concept of plate tectonics suspiciously accommodating. Once one accepts the notion of lithospheric plates sliding under each other like the shuffling of a global deck of cards, almost any geological anomaly has a convenient explanation. Other valid mechanisms for rearranging rocks may get short shrift. As a theory, plate tectonics remains productive and persuasive: as an article of faith, it is sometimes overused.

◄ Continents collide:
1. The Tethys Sea separating the advancing Indian subcontinent from Eurasia narrows.
2. The approaching continental plate begins to push up material from the ocean floor.
3. Compressed beds of sediments are raised into islands.
4. The two plates make contact.
5. The Indo-Australian plate slides under the Asian one, a process that continues 45 million years later.
6. The two continents join along a 600-mile seam, and the sand and seashells of the Tethys Sea are pushed skyward, forming the Himalaya Range.

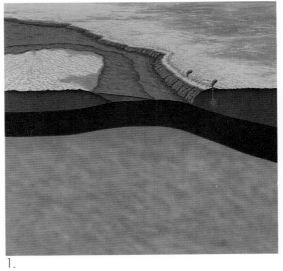

1.

2.

3.

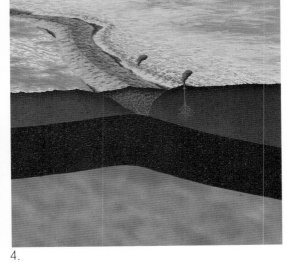

4.

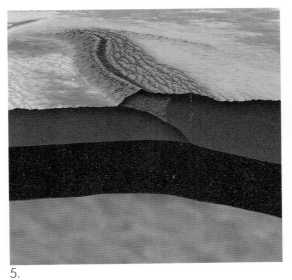

5.

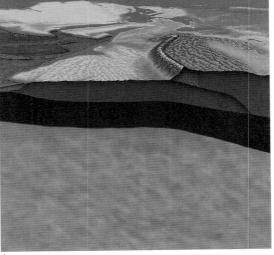

6.

▲ The world's greatest mountain range lies below the ocean, unseen until the advent of research submarines like *Alvin* and *Archimède*. At its summit is a rift valley marking the point where two plates are being forced apart by accumulations of magma.

▲ Pillow lava formed by a lava flow underwater is now visible in the Alps, where marine rocks have been pushed skyward by the forces of plate tectonics.

cold and dark their searchlights revealed numerous spherical rocks on the seabed. Known as pillow lava, these distinctive formations are produced when lava erupts underwater. The characteristic rounded shape results from the abrupt surface cooling as molten rock meets water.

Pillow lava is a familiar phenomenon. It can be seen in shallow water off Hawaii and even in the Alps, where ancient seabeds have been pushed skyward. Its appearance in the deep ocean helped bring about changes in our understanding of undersea geology. During the nineteenth century, most scientists believed that the ocean floor was stable and covered with deep deposits of sediment. More recently, the theory of plate tectonics led their successors to envisage a seascape of fractured geologic plates. Harry Hess, a mineralogist who pursued his ocean floor studies as captain of a World War II attack transport in the South Pacific, hypothesized in 1960 that the ocean floor is "swept clean" by geologic processes every few hundred million years. The minimal sediments found by the 1973 expedition helped confirm Hess's contention, and the new-minted look of the pillow lava at the bottom of the Atlantic near Iceland indicated that this part of the ocean floor is home to active volcanos. Their lava flows are part of a process of constant renewal beneath the sea. Despite increasingly sophisticated searches, the oldest rock found on the ocean floor goes back only about 150 million years. Terrestrial

finds, by comparison, have been dated at over 3 billion years.

In 1977 the *Alvin* was involved in another historic exploration. This time the site was about 125 miles off Mexico's Pacific Coast where, at a depth of 8,500 feet, the researchers discovered a forest of eerie chimney-like structures emitting dark clouds. The billowing discharges turned out to be super-heated water (as much as 600 degrees Fahrenheit) colored by particles of suspended metals. Sea water finds its way underground through cracks in the oceanic plates, is heated by magma and erupts back upward, carrying a load of dissolved minerals from the interior. The chimneys, dubbed "black smokers," are made of solid mineral deposits — primarily iron, zinc, copper sulfides, and silver — precipitated out of solution by the cold water.

The smokers support an incredible abundance of life. Here, in what to us seem almost unimaginably hostile conditions of heat, darkness, pressure, and toxic metals, anaerobic bacteria thrive, providing the basis of a rich food chain. Scientists studying the spreading zone at the Galapagos Rift just north of the Equator found 12-foot-long worms, giant clams, and blind crabs among the array of creatures clustered around vents and smokers.

Smokers are quite common phenomena. Others have been found in the North Pacific, in the Caribbean, and elsewhere. Some, such as the travertine towers of Djibouti, North Africa, are on dry land, but they too originated beneath the water. Terrestrial eruptions are actually a relatively minor component of the Earth's total volcanic activity. The vast majority take place undersea, where the Earth's crust is thinner. Here the great heat generated by radioactive elements in the mantle finds its principal release. Like a giant radiator, a submarine volcanic range runs from the Mexican coast, across the Pacific and Indian oceans, around the Cape of Good Hope, and up the Atlantic Ocean through Iceland, a distance of some 40,000 miles. This deep ocean floor is among the least explored — and little known — portions of the globe. However, the same technology that locates schools of fish and suspicious submarines is being used to map its hills and valleys, and computer models depict a range cradling a valley nearly two miles deep. This rift valley is studded with smokers where volcanic activity is greatest.

Reservoirs in the rock beneath the valley floor periodically fill with magma, water vapor, and gas. The magma heats the rock on either side, some of which also melts. Eventually, the pressure within the reservoir is released in a series of eruptions. The lava that is released hardens to rock again, spreading the sides of the Mid-Atlantic Ridge apart at a rate of almost an inch per year. The Mid-Pacific Ridge is even more active, both in terms of smokers and the rate at which the sea floor is spreading. Laser measurements taken from satellites have helped to confirm this slow but inexorable movement.

Not all undersea eruptions take place in rift valleys. Some of the most visible are the results of isolated "hotspots" where a plume of magma rising from within the mantle has drilled its way through the crust, piling up lava into undersea

▼ Marine mountain ranges rise gradually to long ridges that occasionally, as in Iceland, rise above sea level. Most of the tectonic activity on Earth takes place along these seams.

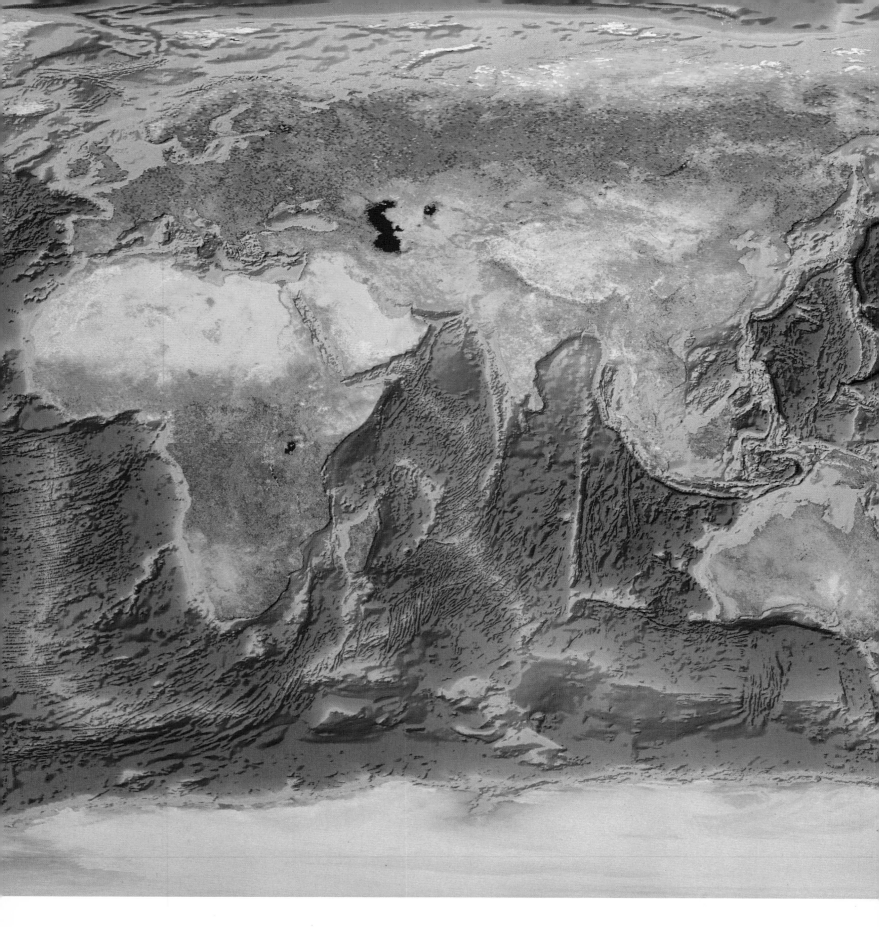

▲ Sonar measurements and first-hand observation have made it possible to map one of Earth's last frontiers, the land beneath the sea. The path of the Mid-Oceanic Ridge is crossed by rows of stress fractures created as the plates are forced apart.

mountains. Hotspots are stationary, but the crust above them is not. As the plate moves above it, the magma plume makes a new hole and then a new mountain. If the plume is active enough, these undersea mountains become emergent islands. Thus the Hawaiian islands of Oahu, Maui, Kauai, and Niihau are older and more eroded creations of the same hotspot that is now pumping lava through Mauna Loa and Kilauea on the more recently created island of Hawaii. An even younger, future

Hawaiian island — already named Loihi — has risen to 3,000 feet beneath the ocean surface south of Hawaii.

CONVERGENCE AND DIVERGENCE

Strings of volcanic islands comprise a sort of diary of Earth's ongoing tectonic history. A comparison with Mars, where evidence of plate tectonics has not been seen, reinforces the role of

Earth's internal heat. Hawaiian mountains like Mauna Loa and Mauna Kea are called shield volcanos because their smooth, gentle silhouettes, built up by a succession of lava flows, resemble the round shields once carried by Greek warriors. Some, like Mauna Kea, have erupted more or less continuously for many years, covering the effects of erosion with new layers of basalt. Similar mountains are found on Mars, and one of them, Olympus Mons, is the largest known volcano in the solar system. Mauna Loa rises six miles above the ocean floor, making it bigger than Mount Everest, but Olympus Mons is 15 miles high and more than 300 miles across the base.

The difference in size stems not from the amount of heat in the two planets. In fact Mars, with its lesser mass, probably never was as intensely hot as Earth. Olympus Mons grew monstrous because on Mars the crust has been stationary for eons, allowing it to sit over the same ·

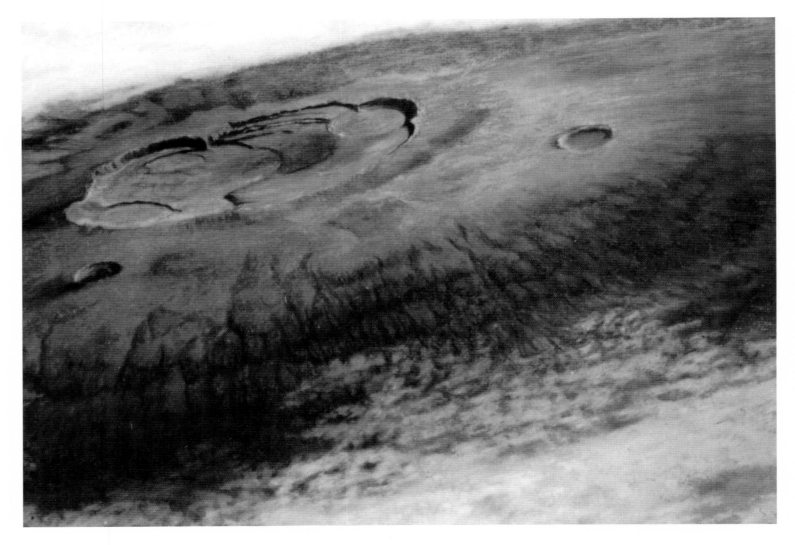

▲ Mars' extinct volcano, Olympus Mons, was photographed by *Viking Orbiter 1*. It is shown in false color which exaggerates very subtle differences in the volcano's lava flows that have accumulated over time. Olympus Mons is the largest known volcano in the solar system, rising about 80,000 feet above Mars' average elevation.

▶ The eruptions and earthquakes that trouble Earth's surface are not isolated events, but consequences of forces at work in its interior.

hotspot. On Earth, Mauna Loa is only one in a series of volcanos created by the same magma plume.

The Earth is not steady beneath our feet. Its crust and upper mantle — collectively termed the lithosphere — are cracked like the shell of a softboiled egg. Some 20 lithospheric plates float on currents of slowly moving mantle. The material

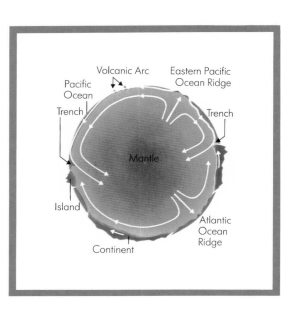

erupted from the rift valleys helps keep the plates in motion.

Rather like the sutures in a baby's skull, the plates help accommodate pressure from below. But unlike a baby's skull, the Earth is growing no larger. Rock that is added along spreading zones such as the one bisecting Iceland must be subtracted from somewhere else. The spreading zones — called divergent plate boundaries — have corollaries in convergent plate boundaries, where the edge of one plate sinks under the edge of another and recycles its rock back into the mantle. This process is occurring now along the west coast of South America, and in an arc from the Aleutian Islands south to the Philippines. Elsewhere, as along the San Andreas fault, parallel plates slide past each other, their periodic shifts and lurches announced as earthquakes.

FOSSILS AND PLATE TECTONICS

The largely undersea activity of plate tectonics is also displayed in the mountains. At Kathmandu, Nepal, nearly a mile above sea level, fossilized seashells are sold as souvenirs in open-air markets. They come not from the coast, 500 miles away, but from the Himalaya Range to the north. The Earth's tallest terrestrial mountain range, with

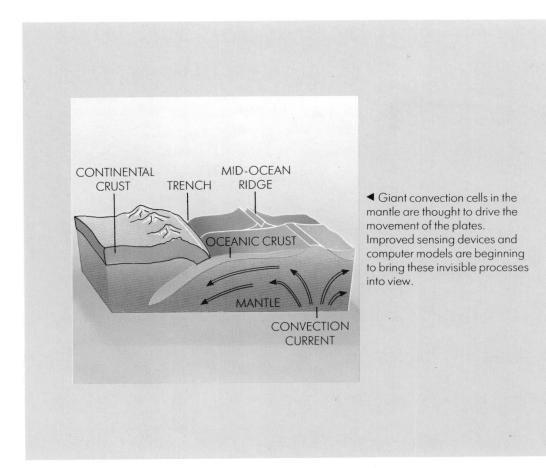

CONTINENTAL CRUST TRENCH MID-OCEAN RIDGE

OCEANIC CRUST

MANTLE

CONVECTION CURRENT

◄ Giant convection cells in the mantle are thought to drive the movement of the plates. Improved sensing devices and computer models are beginning to bring these invisible processes into view.

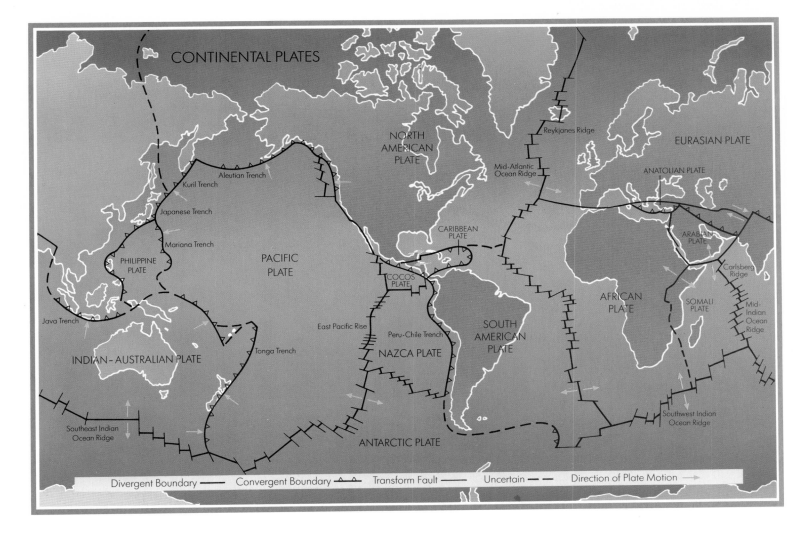

CONTINENTAL PLATES

NORTH AMERICAN PLATE

Reykjanes Ridge

EURASIAN PLATE

Mid-Atlantic Ocean Ridge

ANATOLIAN PLATE

Aleutian Trench

Kuril Trench

Japanese Trench

Mariana Trench

PHILIPPINE PLATE

PACIFIC PLATE

COCOS PLATE

CARIBBEAN PLATE

ARABIAN PLATE

Carlsberg Ridge

AFRICAN PLATE

SOMALI PLATE

Mid-Indian Ocean Ridge

Java Trench

East Pacific Rise

Peru-Chile Trench

SOUTH AMERICAN PLATE

INDIAN–AUSTRALIAN PLATE

Tonga Trench

NAZCA PLATE

Southeast Indian Ocean Ridge

Southwest Indian Ocean Ridge

ANTARCTIC PLATE

Divergent Boundary —— Convergent Boundary —△△— Transform Fault —— Uncertain ---- Direction of Plate Motion ——→

MECHANICS OF MOTION

The prevailing explanation of plate tectonics is that the molten rock in the Earth's mantle circulates in great convection cells, following the same laws of physics that control soup simmering on the stove and air currents above the Equator. The force of these currents propels the plates. Unlike the soup and the air, however, the mantle is not available for direct observation. Geologists must decipher the messages delivered by earthquakes and eruptions to make the assumptions for their computer models.

Until the mid-1980s one of these assumptions was that the convection cells, and therefore the plates, move at a uniform rate. Peter Vogt and John Bronkema, geophysicists at the Naval Research Laboratory in Washington, D.C., have concluded otherwise. In 1985, Vogt told *Science Digest* magazine that the circulation may be more akin to a "lava lamp," with giant blobs of semisolid material rising rapidly through the magma and spreading out against the bottom of the plates. The size and speed of these blobs would presumably cause the plates to accelerate.

Vogt and Bronkema's proposal, like much of plate tectonic theory, is based on measurements of magnetism in rocks. Most rocks contain minerals that align themselves with Earth's magnetic field, creating a record of conditions at the time of their creation. When heated to the temperatures of magma, the minerals lose their magnetic force. Then as the new rock cools, its magnetic field lines up with the prevailing magnetic north and south. So, as Earth's magnetic field reverses periodically, the ocean floor presents alternating bands of magnetically aligned particles.

If the speed of plate movement is constant, the width of the magnetized bands should correspond to the chronology of pole switches established by measurements around the world. It does not. Some bands are wider than the time period would indicate. Vogt suggests the wider bands are the function of periods of faster spreading related to the magma blobs.

Using ocean floor measurements, Vogt plotted the speed of 10 plates over a period of 10 million to 20 million years. All showed the same trends of acceleration and deceleration. These findings may help explain orogenic episodes, the periods of rapid mountain building that have been the subject of much geologic speculation. Vogt hypothesizes that variations in plate speed caused the Himalaya Range to rise in steps, with the fastest periods of uplift occurring when the plates were moving rapidly. Meanwhile, the cycle of eruptions at volcanic hotspots like Hawaii also seems to correspond to periods of rapid movement. The islands of Kauai, Maui, Oahu, and Hawaii all were formed during a period beginning about 5 million years ago, the period that Vogt identifies as a high point in speed.

Establishing variable speed would only send the questions deeper, however. If mantle blobs give the crust a push, what makes the blobs? A possible answer lies in the same laws of convection. New geomagnetic models of the Earth's interior indicate that currents in the outer core bring hot material into contact with cooler portions of the lower mantle. Perhaps the rise of mountains is linked to this variable heat source deep within the Earth.

▶ The rise of the Himalaya peaks has taken place in a series of bursts alternating with periods of slower growth. This uneven development may be caused by blobs of magma circulating in the Earth's mantle.

▶▶ Measurements of the ocean floor help to explain the continuing rise of these Nepali highlands.

14 peaks over 25,000 feet, gives ample documentation of its undersea origins.

Dr. Harutaka Sakai of Japan's Kyushu University has spent years surveying marine fossils in the Himalaya Range. He and his team have catalogued a veritable aquarium of the Mesozoic era. Delicate crinoids (sea lilies), relatives of modern sea urchins and starfish, are found in a cliff more than 10,000 feet above sea level. Ammonites, belemnites, corals, and plankton are all found as fossils in the mountain rock. They represent species that lived in the shallow Tethys sea during the Cretaceous period, and were eliminated some 65 million years ago in the same mass extinction that wiped out the dinosaurs.

At 6,500 feet the geologists found a fossil impression left by the sea itself. Its rippling rock surface matches the pattern left in the sand by waves in shallow water. Even the summit of Mount Everest displays yellow bands of limestone that were formed underwater out of the remains of countless marine organisms.

Most of the fossils are not new discoveries. Early in the nineteenth century, their presence in these mountains was cited as proof of Noah's flood, but a more persuasive explanation lies in the processes of plate tectonics. High-altitude seashells are another result of the same oceanic spreading that is still creating Iceland. To understand the birth of the Himalaya Range, we need to travel back to Earth's beginnings.

CONTINENTAL DRIFT

Continents formed gradually, congealing on the surface of the magma as the protoearth cooled. As the crust solidified, their motion slowed but did not stop.

During the Paleozoic era, about 280 million years ago, the landmasses joined into one giant continent called Pangaea, or "all lands." Under this immense, insulating blanket of thick continental rock, magma continued to build up under hotspots. Eventually, the thermal forces began to

▲ Fossils of ammonites that lived 50 million years ago are sold along with Nepali silver and turquoise in Kathmandu.

◄ The yellow bands on these Himalayan summits are layers of limestone from the Tethys Sea. Their distinctive color serves as a warning to mountain climbers, who must use extra caution crossing the crumbling sediments.

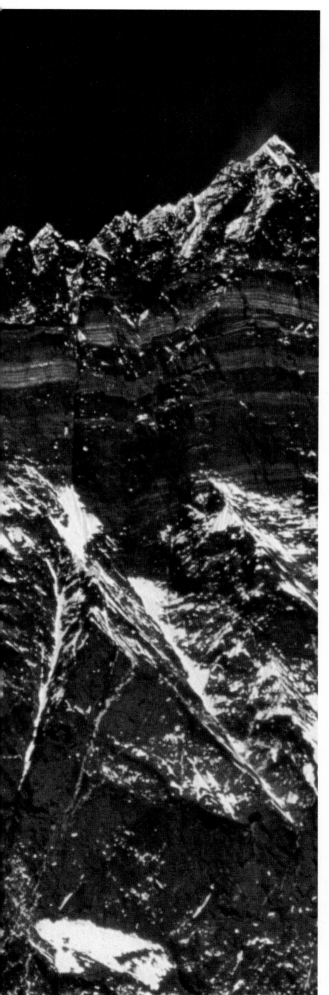

▲ A hundred million years ago, this Himalayan plateau was a forested coastal plain.

split the landscape into new configurations. One section, the Indian plate, broke loose from what is now Africa and Antarctica and was carried north as the prow of a lithospheric plate. Five thousand miles and perhaps a hundred million years later, it reached Asia. In a literally earthshaking collision, the Indian plate ground its way under the Eurasian plate, and the two continental masses joined. The Tethys Sea that had separated the continents was

gone, its floor compressed and folded by the pressure of the advancing plate and then pushed five miles into the sky. The point of collision is still visible as a line running for 600 miles along the base of the Himalaya. The Karakoram Range and the Hindu Kush are results of the same collision.

Satellite measurements confirm that the process is still underway, though proceeding at about half the pre-collision rate of some 30 feet a century.

◄ The meeting place of continents is here along the Kali Gandak river in Nepal. Already over 12,000 feet, the plateau is being pushed even higher by the advancing Indian plate far beneath. The crust shifts and cracks under the strain of this movement, causing violent earthquakes as far away as central China.

▲ The ramshorn shape that gives ammonites their scientific name (from the ram-headed Egyptian god Amon) is clearly visible in this well-preserved fossil from the Japanese island of Hokkaido. It is now on display in the Mikasa Municipal Museum.

◄ During the Cretaceous period, the warm, shallow waters of the Tethys Seas supported a varied ecology including ammonites, belemnites, and the marine dinosaurs that fed upon them.

57

India has pushed some 1,200 miles laterally under the Eurasian landmass creating the thickest continental crust on the planet, and the mountains are still rising. The stresses created by the continuing movement were brutally apparent in the earthquake that hit Nepal and northern India in August 1988.

Farther west, the collision of the African plate and the European landmass some 80 million years ago was responsible for the Alps. The sea that lay between the European and African plates when Pangaea began to separate was more than 500 miles wide, and the compression of its bed into a width of only 60 miles resulted in the dramatically folded rockfaces of the Alps.

MINERAL CONCENTRATIONS

The movement of lithospheric plates has created much of today's geology. It also is responsible for much of our mineral wealth. The Andes Mountains in South America reveal the link between plate movement, mountain building, and metallic ore deposits. The Andes are one of the world's major mountain ranges, spanning 5,000 miles and including several peaks over 20,000 feet. They were created when the Nazca plate below the Pacific Ocean moved beneath the South American continental plate. This process sends a cold oceanic crust (including the remains of smokers, where some minerals have already been concentrated) and large amounts of seawater to a depth of 100 miles below the Earth's surface, where the ancient sea floor melts into magma. The newly molten material collects in underground reservoirs, and, as pressure builds up, magma and steam push their way along fissures, carrying dissolved minerals in solution. When the magma cools, either within the Earth or on the surface after a volcanic eruption, minerals crystalize out — copper, gold, silver, and tin.

Chuquicamata, the largest known deposit of copper ore in the world, is on the northern edge of

▲ Some ammonites reached enormous size.

▲ Ammonites are also found in the Alps.

◀ Cho Oyu, on the border between Tibet and China, rises more than 25,000 feet above sea level, only to be dwarfed by neighboring Mount Everest.

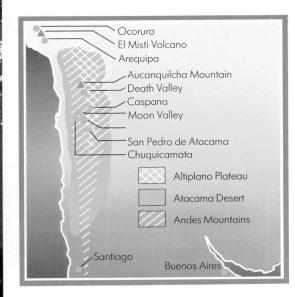

Ocoruro
El Misti Volcano
Arequipa
Aucanquilcha Mountain
Death Valley
Caspana
Moon Valley
San Pedro de Atacama
Chuquicamata

Altiplano Plateau

Atacama Desert

Andes Mountains

Santiago

Buenos Aires

◄ Over its first billion years, Earth developed a series of insulating layers that have helped regulate and maintain its internal heat. The molten surface of its early years cooled over the more intensely radioactive interior. Above the surface, outgassing created a secondary shield of atmosphere that now moderates the flow of radiation to and from the planet.

▲ Limestone formed in the Tethys Sea now tops a hill in southern France. When the pressure of compression is great enough, limestone metamorphoses into marble.

▶ A waterfall traces a path through the twisted Alpine rock near Annecy, France. As the African continent moved north, marine sediments were pushed before it, creating one of the world's most geologically complex mountain ranges.

▶▶ Folding was sometimes so extensive that rock layers were completely doubled over. Limestone in the Alps near Grenoble was forced into an Omega shape. Since then erosion has consumed the summit, leaving a row of upended strata.

the Atacama Desert in Chile. Mines have been in operation there since the seventeenth century. Before the arrival of a narrow-gauge railway, llamas were used to pack ore to the coast, 90 miles away and 9,000 feet down.

Another fabled site of copper is the island of Cyprus, where mining began 6,000 years ago and continues today. Copper inspired the technological advances of early Mediterranean civilization and fueled the first of many territorial disputes over Cyprus.

COPPER: AN ABUNDANT METAL

The copper on Cyprus was concentrated by smokers in an undersea spreading zone that was later thrust above sea level by the convergence of the African and Eurasian plates. The pillow lava and other volcanic features of the Troödos Massif, where most of the ore is found, corresponds closely with what is known about the composition of oceanic plates.

The geological processes that drive plate tectonics do not create elements. Rather they can concentrate and relocate them. Copper, one of the Earth's more abundant metals, still makes up only

◄ Greenish copper deposits are visible on the surface of the Troödos Massif.

0.0058 percent of the crust. It must be concentrated 80 to 100 times before mining is economical. Mining enterprises have been quick to use the theory of plate tectonics as an aid in locating valuable deposits. In turn, their surveys provide more data for testing the theory.

◄◄Copper deposits on the island of Cyprus were concentrated enough to mine and smelt with the technology available 3,000 years ago. Some of the original mines remained in production until recent times.

SPLITTING AND SPREADING

Reports from space have also contributed to the understanding of processes so vast they tend to overwhelm the earthbound imagination. Harrison Schmitt, a geologist and an American astronaut in

◀ An Alpine rockface near Grenoble.

▼ Pressure, heat, and erosion all leave records that have become decipherable once the basics of plate tectonics are understood.

1972, told Mission Control that the view from Apollo 17 "could make a believer out of anybody." The source of Schmitt's epiphany about plate tectonics was the Afar Triangle of Northern Africa, where spreading zones are separating the adjacent Arabian, Somali, and African plates. On one side of the triangle, the Red Sea separates the Arabian Peninsula from Africa. From space it takes only a little imagination to conclude that the two landmasses once were joined, and events on land indicate that the split is growing.

Djibouti, North Africa, like Iceland, is built on lava. Unlike Iceland, its daytime temperatures commonly exceed 100 degrees Fahrenheit. The land is dry, but beneath the surface, water is at work: the travertine columns that dot the desert emit periodic bursts of steam. These turreted formations, some 200 feet tall and still growing,

◀ Copper ore from the mines at Chuquicamata is concentrated by heating to remove impurities.

◀◀ Copper deposits at Chuquicamata start at the surface and descend several miles into the ground.

◄ The eastern arm of the Great Rift Valley is dry, but if geologists are correct it may eventually be an inlet of the African Sea.

◄◄ Parts of Djibouti go for years without rainfall, but water from the Red Sea moves along underground cracks, and surfaces, steaming, in the turreted landscape.

probably began as smokers on the bottom of an ancient lake. Today they mark a fissure line where the earth is sinking, splitting, and spreading. Seawater seeps into the fissure, is heated by the magma beneath, and rises back to the surface as steam.

This part of Africa is riding a geological express train. Laser measurements show that every year the valley of the smokers is widening by over half an inch. The process is announced by daily earthquakes whose epicenters stretch along the Assal Rift between the Gulf of Aden and Lake Assal. The whole area is sinking with the quakes, creating

an ever-deeper, terraced valley that marks the trench between two plates. At Lake Assal, encroaching seawater arrives via underground springs, then evaporates in the desert air, leaving a thick rim of salt around the shore.

Further south, this splitting is taking place on an even larger scale. It is evident in Africa's Great Rift

▲ Geologists from the Institute de Physique du Globe at the University of Paris use laser beams to measure the spreading of a valley in Djibouti.

◄ The Great Rift Valley. Twenty million years ago a series of volcanic domes along a plate boundary underlying Africa were formed. The stresses of expansion eventually split the domes along their tops, forming the valley.

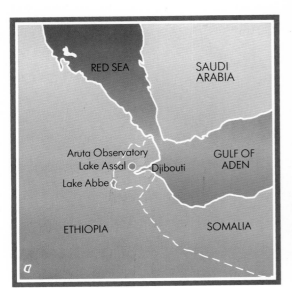

RED SEA
SAUDI ARABIA
Aruta Observatory
Lake Assal
Djibouti
GULF OF ADEN
Lake Abbe
ETHIOPIA
SOMALIA

EARTHQUAKES

Continents move slowly. Although at the top of Mount Everest we can see limestone orginating from the sea floor, we cannot tell with our own eyes that the summit is still rising. Earthquakes and volcanos, however, bring plate tectonics into a human time frame. Mountains rise from fields and cities are reduced to rubble at an appalling speed.

Along the San Andreas fault in California, the Pacific plate is sliding north past the North American plate at the rate of about two inches a year. Such a plate boundary is called a transform fault margin. The movement is visible in the California town of Hollister, where sidewalks and foundations straddling the fault line periodically crumble and split.

Despite the insecurities of life in Hollister, earthquake experts are more worried about San Francisco and Los Angeles, the metropolises to the north and south. There the fault is not moving freely and the strain is building up. Some day — and seismological evidence suggests it may be soon — the Earth will lurch into a new arrangement. The relatively modest quakes that have shaken both cities in recent years are not enough to relieve the stresses beneath them.

Transform faults are not the only source of earthquakes. Along subduction zones, where one plate is forced beneath another, they can be even stronger. The San Francisco quake of 1906 is estimated to have measured 8.3 on the Richter scale, where each scale point represents an order of magnitude. Quakes in deep ocean trenches along subduction zones have been measured at 8.9, six times as intense. The 1960 earthquake in Chile measured 9.5. Satellite observations of fault systems indicate that the quakes that flattened Tangshan, China, in 1976 were a response to the plate movements that are still building the Himalaya Range more than 1,000 miles to the southwest. More than 200,000 people were killed in the Tangshan quake, which was by no means the most deadly in Chinese history.

With a huge territory to cover and limited resources, China's seismologists have mobilized the population to watch for signs of something coming. Farm animals are known to act strangely when a quake is imminent, and as well, catfish are kept in tanks and any sudden agitation is noted. Farmers monitor irrigation canals for unusual turbulence or odor.

On the other end of the technology spectrum, scientists bounce laser beams off fixed-orbit satellites to measure movements of the plates. Still more information is expected from the Earth Observing System, a joint effort of the United States, Japan, and the eleven-nation European Space Agency that is planned for the end of this century. The project will include four orbiting space platforms, and the combined data from different vantage points will help observers analyze plate movement as a related system rather than a series of disparate events. This same cosmic perspective is being applied to the layers that form the planet. Already, measurements derived from radio emissions of the farthest stars are being studied to see if they can pick up surface swellings produced by fluid moving in the Earth's core. If the data are precise enough, scientists will be using radio waves from billions of light years away to measure events 6,000 miles beneath their feet.

◄ The terraces of the Assal Rift Valley in Djibouti are created by a deepening fissure at the site of a divergent plate boundary. Daily earthquakes by the dozen announce the movement taking place underground.

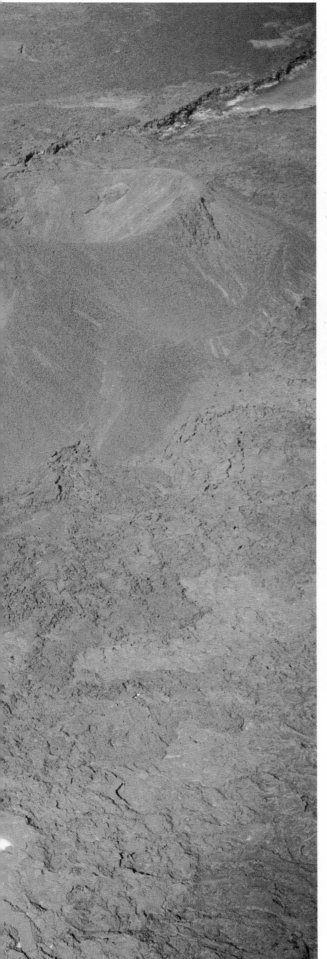

◄ Villagers in Eburru, Kenya, collect water condensed from steam that rises from a fissure in the earth. As in Iceland, underground water is heated as it moves past a magma chamber.

▲ Women congregate in the bazaar of Djibouti's namesake capital city. Many vendors sell salt evaporated from the encroaching seawater, one of the country's few export products.

SUDAN

ETHIOPIA

Lake Rudolf

UGANDA

▲ Teleki Volcanic Arc

KENYA

Lake Baringo

Elemtosta

Lake Victoria

Lake Magadi

Herod's Gate

Logonoto Volcano

Lake Natron

Ol Doinyo Lengai

TANZANIA

INDIAN OCEAN

Valley, which stretches from the Red Sea to Mozambique and on a clear day is visible from the Moon. Like the mid-oceanic ridges, the valley tops a mountain range built up by volcanic action. A great chain of lakes, their water trapped by the sinking earth, runs through Kenya, Rwanda, Burundi, and Tanzania.

The splitting process is slower here than in the Afar Triangle, but the activity in fissures is evident. In Eburru, Kenya, villagers collect water by condensing steam from an underground spring. If the spreading continues the Rift will keep widening. In 50 million years the continent will be split in pieces and the global map may show an African Sea.

◄◄ Lake Assal is more than 500 feet below sea level. Only the lava flows of the Assal Rift separate it from the Gulf of Aden seven miles away. Sea water percolates through cracks in the rift floor to reach the lake, where evaporation leaves great crusts of salt along the shore.

◄ Colonies of stromatolites thrive in the salty waters of Hamelin Pool, Australia. The stony cushions are made by the descendents of the primitive bacteria that first gave the Earth free oxygen.

3

LIFE FROM THE SEA

BECAUSE we belong to an essentially parochial, terrestrial species, we name our world Earth when its essential attribute is water. Oceans are the predominant feature of the planet, covering more than 70 percent of its surface. Water is the agent of creation and erosion, the primary constituent of every living thing we know. Its rhythms are in our cells, necessary and inescapable.

Molecules of water were among the original compounds in the solar system. They were dispersed in meteorites to all of the planets, but only Earth has liquid water today. This crucial distinction is the result of size and distance. Earth receives the right amount of sunlight to keep its median temperature at about 60 degrees Fahrenheit, comfortably between freezing and boiling. It was not always so.

The Earth's first atmosphere came from the same source as its solid rock — that is, the vast cloud of matter from which the solar system was formed. While the young Earth was a still-accreting rocky ball, it was surrounded by a layer of the most prevalent gases in the solar system, including helium, argon, and neon. These are among the "noble gases," so named because they are relatively inert, seldom forming compounds with other elements. Their rarity in Earth's present atmosphere is a result of the development of the protosun into a functioning star. When the Sun accumulated enough mass to begin thermonuclear reactions in its interior, the result was a shock wave of matter and radiation that stripped the primordial atmosphere from all the inner planets.

The air and water the Earth produced henceforth would come from its own ground.

Under Earth's cooling crust, rocks melted again in the radioactive ferment of the mantle, liberating water vapor and other gases that were sent skyward in volcanic eruptions, a process called outgassing. Within as little as 100 million years, volcanic activity had produced a new atmosphere high in water vapor, carbon dioxide, nitrogen, and hydrogen compounds.

When temperatures cooled enough, water vapor began to return to the surface as rain. Rocks found at Isua, Greenland, indicate that oceans were

◀ A stromatolite column in Western Australia. More than three feet high, this structure is the work of thousands of generations of bacteria.

◀◀ The Great Barrier Reef, one of Earth's most complex and lovely environments, is also a link in the carbon cycle that helps regulate global climate.

▶▶The shores of Great Slave Lake in the North West Territories of Canada are lined with fossilized stromatolites, the legacy of vast colonies of cyanobacteria that flourished there two billion years ago.

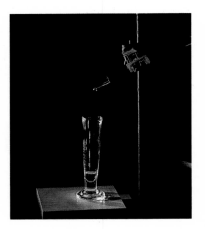

▲ A simple experiment extracts water from a pulverized section of the Murchison meteorite. The laboratory demonstration is a microcosm of the process that eventually covered most of Earth with water.

▶ Fragments of the Murchison meteorite at the Field Museum in Chicago. Chemists have identified a number of the organic compounds and amino acids common to living things in the meteorite. Some of these organic chemcials are able to assemble themselves into a fluid surrounded by a membrane, a non-biological analog to the earliest living cells.

present within a billion years of the planet's formation. Dated at 3.7 billion years, the iron-rich sedimentary rocks were formed at the bottom of an ancient sea. By two billion years ago, Earth had the same volume of water it has today. Its warm, shallow oceans might have been unrecognizable to a modern geographer, but their composition was surprisingly similar to modern waters.

On the newly solidified continents, lakes trapped rain and rivers transported minerals from eroding land to sea. Supplied by terrestrial weathering and undersea eruptions, oceans carried a varied chemical brew. Simple compounds began to combine into the complex strings of molecules that are the structural components of life. Some of these organic chemicals have been identified in a meteorite that struck Australia in 1969. The meteorite, named Murchison after the site where it landed, was probably part of an asteroid formed at the same time as Earth. It also contains impressive amounts of water trapped in its minerals.

Organic chemicals also have been synthesized in the laboratory by exposing an approximation of Earth's early atmosphere and ocean to an electrical charge. Occasionally, conditions in nature also approximate what might have been found four billion years ago. During the eruptions that created the Icelandic island of Surtsey, for example, lightning flickered over a mixture of water vapor, sulfur dioxide, and carbon and hydrogen compounds, all suddenly heated to boiling point by the lava. When the smoke cleared, scientists from the National Aeronautics and Space Administration found organic compounds that did

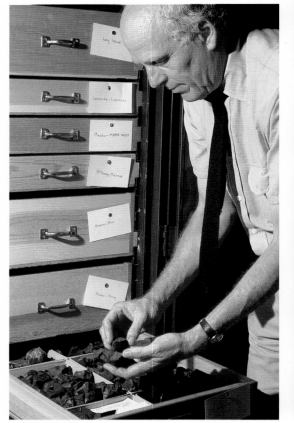

▲ Terrace of the Goddess in the Mammoth Hot Springs area of Yellowstone National Park.

not match those present in the surrounding seawater. Although the possibility of contamination makes such field experiments difficult to verify, the researchers think the chemicals were synthesized during the eruption.

Its exact composition is unknown, but Earth's atmosphere at the dawn of life is thought to have resembled those still found on Mars and Venus,

high in carbon dioxide with only the smallest trace of free oxygen. This lack was providential. Without the mediation of living enzymes, organic compounds cannot develop or survive in the presence of oxygen. The same oxidizing process that turns iron into rust breaks down their bonds. Until cells developed defences against oxidation, what is now the breath of life was a lethal poison.

WATER: THE CATALYST OF LIFE

The transformation from chemicals to living things was made in water. Perhaps the crystal structure of clay or another mineral served as a template on which organic molecules could organize themselves into a more complex structure that then was animated by a bolt of lightning.

Perhaps the heat and abundant nutrients around ocean smokers gave the organic compounds the boost they needed. One recent, and disputed, computer analysis traces the ancestral gene of all life on Earth to a form of bacteria that grows in a mixture of boiling water and traces of sulfuric acid. Similar organisms can still be found in volcanic areas like Mount St. Helens and Yellowstone,

▶▶Newly deposited limestone at Mammoth Hot Springs is gray or white. Browns, yellows, and other colors are the creation of colonies of algae.

▲ A mixture of steam and hot gases bubbles from the pools and mudholes of Yellowstone. Although toxic to most life, the odoriferous mudpots are home to several species of bacteria.

▶ The cyanobacteria that make stromatolites, shown under magnification.

where their brilliantly colored growth contributes to sights like Grand Prismatic Pool and Artists Paintpots. However, opponents of the boiling bacteria theory point out that high temperatures are as devastating as free oxygen to organic chemicals. These scientists favor lakes or lagoons as the first sites of life.

Wherever they originated, the first organisms were probably little more than particularly complex organic molecules surrounded by a membrane. Their internal organization was not sophisticated enough to let them synthesize their own food out of simple molecules. Instead their nourishment came premixed from the chemical smorgasbord around them. Lacking a nucleus, they were incapable of sexual reproduction, but they did contain genetic material. They could multiply and they could evolve.

The next three billion years, 85 percent of the tenure of life on Earth, were presided over by an ever-increasing variety of single-cell organisms. Levels of radioactivity and ultraviolet radiation were higher in the first eons of life on Earth, and mutations were probably frequent. Most were failures, but some allowed organisms to stake out new niches. The cells divided and conquered, colonizing the waters and transforming the planet to a degree none of their larger descendants has

PHOTOSYNTHESIS

equaled. From the atmosphere above us to the limestone beneath our feet, much of the world owes its existence to this microscopic life.

As these first organisms prospered, they began to deplete the supply of oceanic "fast food," the carbon compounds that had accumulated through millions of years of chemical reactions. The evolutionary advantage went to organisms that did not require premixed chemicals but could synthesize complex nutrients out of widely available molecules. One family of bacteria developed a mechanism to get nutrition from carbon dioxide and hydrogen, producing methane as a byproduct. These methanogens still thrive in the stomachs of ruminant animals, fueling the chemical reactions that make a steer an efficient converter of hay into sirloin.

Then, perhaps as much as four billion years ago, bacteria produced an organic molecule with the

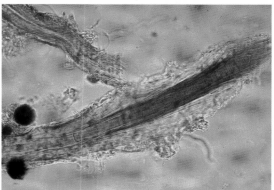

property, unique in living tissue, of being able to store the energy in sunlight. This green-tinted substance is now called chlorophyll and the organisms that pioneered its use are known as cyanobacteria. Supplied with chlorophyll and sunshine, cells became alchemists. Photosynthesis enabled them to create high-quality carbohydrates from the abundant raw materials of carbon dioxide and hydrogen.

▲ Primitive organisms are not the only life to thrive in the special environment of Yellowstone. Coyotes and bears find heated dens in the Mammoth area, while a hot-climate lizard, *Sceloporous graciosus graciosus*, has retained a small niche further south. A low-growing grass hugs the warm ground, staying green even in subzero temperature.

◀ Even greater magnification of a stromatolite bacterium shows the gelatinous sheath that surrounds the cell wall. The dark spheres are grains of sand trapped by the sticky surface.

AEROBICS

Photosynthesis set the stage for one of the great dramas of Earth's history, one which is no less stirring for the fact that its stars were single-celled organisms. The addition of free oxygen into the atmosphere was a major crisis for life on Earth, but in the tradition of inspirational stories, primitive bacteria found a way to turn danger into opportunity.

From a cell's point of view, free oxygen is a good friend but a bad enemy. It is extremely reactive, combining easily with organic compounds. These reactions can release a great deal of energy. Oxygen metabolism contributed to the tremendous evolutionary burst that began 600 million years ago and gave the Earth its present diversity of creatures. The same reactive qualities, however, make oxygen destructive to unprotected cells. Lacking some kind of shield from its reactive effects, early organisms found even minute quantities fatal. Some of their descendants cannot tolerate even one percent of current oxygen levels.

As oxygen levels rose, intolerant species either died or found protected niches in sediments and volcanic vents. Later they took advantage of the new world made possible by oxygen, finding homes in waterlogged soil and, by the trillions, in the intestinal systems of animals. These obligate anaerobes are as dependent on the absence of oxygen as humans are on its presence.

More adaptable organisms evolved to tolerate low levels of oxygen, colonizing transitional areas like mudflats and stagnant ponds. Still others can handle a range of oxygen levels, switching from anaerobic to aerobic metabolism to fit the situation at hand.

Inside the cells, the key to oxygen tolerance lies in enzymes. These specialized chemicals are the mediators of chemical reactions, providing the mechanism for the release of energy while protecting the essential chemical bonds of the cell. All cells of higher organisms contain enzymes for oxygen metabolism. Their presence was a prerequisite for the development of multicellular life.

Photosynthesis also affected Earth's climate. The burgeoning stromatolite colonies at Great Slave Lake may have contributed to the first known ice age 2.3 billion years ago. By depleting the carbon dioxide in the atmosphere and ocean, they thinned the insulating blanket that kept the planet warm despite a dimmer Sun. The oxygen they produced in exchange was much less efficient at holding heat. The same cycle may have helped bring on the end of that episode. If cold weather or a shortage of carbon dioxide lowered the stromatolite populations, oxygen production would diminish. Without as many organisms to consume it, the carbon dioxide released from sediments would re-enter the atmosphere and allow the planet to warm again.

Carbon dioxide levels are once again on the rise. The destruction of coral reefs, the burning of coal and petroleum, and the logging of rainforests are weakening links in the carbon cycle. More of the gas is added to the atmosphere and fewer organisms remain to soak it up. If the trend continues it will once again alter the world that life has helped to fashion. This time humans as well as bacteria will have to rise to the occasion.

◄ The oxygen-poor waters of the Okefenokee Swamp in Georgia and northern Florida are a haven for organisms that cannot tolerate modern levels of atmospheric oxygen.

Photosynthesis was a tremendous evolutionary breakthrough, but it had one major drawback. Its byproduct, oxygen, was deadly to the organisms that created it. Until cells developed enzymes to protect their internal structure, they were in mortal danger. Many species did die, and others went into hiding.

ATMOSPHERIC OXYGEN

The first biologically produced oxygen was drawn into chemical reactions in the ocean. Iron rusted into ferric oxide; carbon oxidized into more carbon dioxide. Hydrogen joined with oxygen to make more water. The introduction of oxygen can be traced by examining iron deposits in rocks of various ages. Banded iron formations, found in rocks over two billion years old, show a puzzling combination of ferrous and ferric iron. They may

◄ The ancient rock at North Pole, in northwestern Australia, has been subjected to a minimum of heat and folding since its formation more than three billion years ago. Both the rock and the fossils it carries are surprisingly well preserved.

▼ The earliest fossils at North Pole are found in a layer of chert, a grayish rock made of microscopic grains of silica precipitated from ocean water. Ancient oceans were much richer in dissolved silica because the sponges and diatoms that now extract it did not yet exist. The glowing red color now seen at North Pole comes from iron-rich surface deposits.

be from a period when free oxygen was present in localized areas.

Eventually the oxygen carried in water reached saturation point, and the element made its appearance in the atmosphere. "Red beds" of oxidized iron show up in rock younger than two billion years, and demonstrate the effect of atmospheric oxygen on terrestrial rock. By a billion years ago, oxygen probably approached its modern concentration of 21 percent of the atmosphere. The rise of oxygen can be traced in surviving species of cyanobacteria. Several kinds are able to switch to anaerobic synthesis in the absence of oxygen, and many others grow best at lower than modern concentrations. At present, photosynthesis releases some 20 billion tons of oxygen per year. This production is balanced by the cycles that consume oxygen, among them the metabolism of humans and other air-breathing animals, and the oxygen consumed in fires.

Many of the organisms that produced this change were present within a billion years of Earth's formation. By then the original thread of life had branched into a complex tapestry. In 1976, rock samples taken from the desert near a place whimsically called North Pole, in Western Australia, contain a variety of organisms, including cyanobacteria dated at 3.5 billion years. Though still confined to one-celled organisms, the North Pole fossils show evidence of considerable evolution.

◄ The limestone cliffs of the Napier Range in Western Australia are a fossilized coral reef formed 350 million years ago. Coral reefs are complex associations of plants and animals and are the Earth's oldest ecosystems, with examples going back nearly 600 million years.

▼ A section of a fossilized stromatolite at North Pole. The structure is the same as in the living examples at Hamelin Pool not far away.

91

▲ Unlike stromatolites, which are the product of cyanobacteria alone, coral reefs are an association of both plants and animals. Functioning like the sea anemones they resemble, these corals sweep the waters for microscopic prey. Some of the largest species even capture fish.

▶ Stromatolites at Hamelin Pool. Gentle waves keep the minute builders supplied with sand and debris.

Living relics of this period are also found in Australia. At Shark Bay on the west coast, Hamelin Pool is so salty that little life can endure it. Its calm, shallow water is studded with formations that resemble large stepping stones. Although their name, stromatolites, comes from the Greek for "bed of rocks," they are in fact a collection of cyanobacteria. The stepping stones are their creation.

Stromatolites were first identified in fossils from New York state in 1825, but it was not until 1954 that living examples were discovered in Shark Bay. Since then they have been found in about 20 spots around the world, ranging from Great Slave Lake in northern Canada to the Persian Gulf. Adaptable to

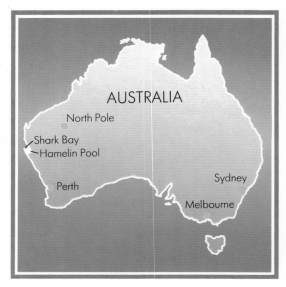

AUSTRALIA

North Pole

Shark Bay
Hamelin Pool

Perth

Sydney

Melbourne

▲ The Great Barrier Reef is part of the largest biological construction project on Earth. Spurs projecting from its outer margins deflect and dissipate the force of incoming waves, protecting the shorelines and fragile islands within its boundaries.

many environments but vulnerable to predators, they flourish where few other creatures can. Fossil stromatolites are found in the same strata that contain the oldest North Pole fossils; they rank among the oldest known life forms.

STROMATOLITES AND OXYGEN

The organisms that create stromatolites group together into a soft, slimy mat. Viewed under a microscope, an individual bacterium emerges as a long, thin cell covered by a sticky sheath. The sheath catches sand and other debris carried by the current. During the day, the colony uses sunlight to synthesize glucose and calcium carbonate from water and carbon dioxide. At night it secretes calcium carbonate, which mixes with the trapped debris into a sort of cement which hardens into a layer of rock. Some of the bacteria are trapped and die inside their own creation, but some survive to form a new covering mat and to continue the process. The result is a layered stone sandwich, growing at the rate of about half an inch a decade.

Once these organisms were the planet's dominant life form and its major source of oxygen. The fossil record they left also provides clues to the world they came from. At Great Slave Lake, fossilized stromatolites nearly 100 feet thick stretch along 60 miles of shoreline. Their striations result from alternations of dark and light, warm and cold,

STROMATOLITE CONSTRUCTION

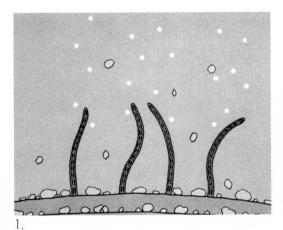

1.

2.

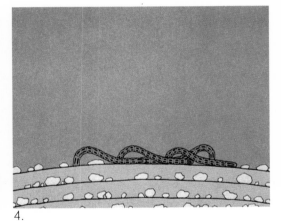

3.

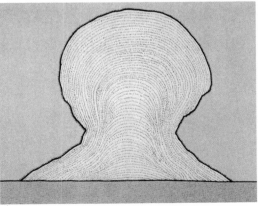

4.

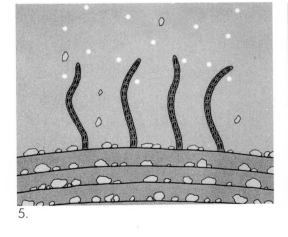

5.

6.

◄ Building rocks from sand and sunlight:

1. The cyanobacteria that form stromatolites are active during the day, absorbing sunlight and picking up fragments of sand and debris on their sticky sheaths. the white bubbles represent oxygen released during photosynthesis, a byproduct that transformed Earth's environment.

2. At night, when the organisms are quiescent, the sheaths and debris meld into a thin paste.

3-4. Even as sunlight dries and hardens the previous layer, surviving bacteria on the surface begin the process again.

5. The column's shape is influenced by a number of factors including day length, sediment patterns, and surf.

6. After many thousands of repetitions, the bacteria colony's daily cycle is expressed in stone.

indicating a 435-day year. Similar cycles have also been found in coral, giving credence to calculations that the Earth's rotation speed has decreased over time. As the spin slows, days get longer and fewer fit into a year.

CARBON TRANSFORMED

Photosynthesizers like those that formed stromatolites did more than introduce oxygen into the existing atmosphere. By withdrawing calcium carbonate from the sea and locking it up in the rock, they began to reduce the levels of atmospheric carbon dioxide. At the time stromatolites flourished around Great Slave Lake, some 2.5 billion years ago, carbon dioxide may have been more prevalent in the atmosphere than oxygen. Today its atmospheric level is about 0.03 percent, and even small rises in that level create international alarm. The carbon-dioxide-rich atmosphere that helped

▼ The rough stone highlands of Northwest Australia still show their origins as coral reefs.

warm the young Earth before the Sun attained its current heat would cook the planet now.

Most of the carbon dioxide contained in Earth's primitive atmosphere is now moving through some other part of the planet's carbon cycles. Trees, shells, and ocean sediments are all part of the never-ending transformation of carbon from one form to another. The work of time and shifting continents has transformed a simple molecule into works of wonder.

◄ Fossilized coral is one indication of the genesis of the limestone rock that now covers nearly a quarter of the land surface of China.

GAIA

Since life began, Earth and the creatures upon it have shared an intricate relationship. The creation of atmospheric oxygen is only one of the ways in which living things have altered the very conditions that make life possible.

In a controversial hypothesis, James Lovelock and Lynn Margulis have extended this understanding to include the proposition that Earth can best be understood as a living being, actively regulating her chemical cycles to maintain conditions suitable for life.

Lovelock, a British chemist, named his hypothesis Gaia in honor of *Ge,* a Greek Earth goddess also commemorated by the words geography and geology. A pioneer of geology, James Hutton, had suggested 200 years earlier that the Earth functions as an organism and its cycles are analogous to the circulation of blood in a living body.

One of Lovelock's arguments for Gaia is the stability of Earth's climate. Despite major changes in atmospheric gases and solar radiation, temperatures have stayed within the relatively narrow range compatible with life. Climatic changes have caused extinctions of specific species, but for more than three billion years, most of the planet has remained a temperate haven. This seems all the more remarkable considering that even a brief period of lethal global temperatures would probably have irreversible effects. Earth would no longer meet the conditions that allowed life to evolve in the first place.

The prevailing scientific view is that climatic equilibrium is a lucky accident. Lovelock and Margulis, a Harvard biologist, contend that it is an automatic adjustment on the part of a complex living system, more like shivering or sweating. Much of their thesis rests on the composition of the atmosphere, which Lovelock calls "a highly improbable distribution of molecules." Its relative abundance of oxygen and methane and scarcity of carbon dioxide could not be maintained without biological cycles. Without life, they say, the atmosphere would find a new equilibrium based on surface temperatures in the range of 500 degrees Fahrenheit.

Extending Hutton's analogy, the Gaia hypothesis assumes that Earth has some vital organs and some expendable ones. Just as a human can live without a hand but not without a heart, Gaia might be able to sustain devastation in its temperate landmasses (however much the individual citizens of those regions might object) but not the destruction of estuaries and tropical forests.

Lovelock places chemical cycles and Darwinian evolution at the center of his exploration of Gaia. He gives the planet a purpose — the continuation of life — but does not embue it with magical means to achieve its goal. Nevertheless, his theory has been dismissed by many of his colleagues as a mystic intrusion in the halls of science. Yet he is by no means the only scientist to pursue the broader implications of life and consciousness. The ambiguities of quantum mechanics, in which the behavior of subatomic particles can be described only in terms of probabilities, has inspired a number of meditations on the nature of reality.

In *The Symbiotic Universe,* George Greenstein concludes that life is no random by-product of universal forces but rather its *raison d'être.* Given the uncertainties of the smallest components of physical reality, Greenstein concludes that nothing can be said to exist unless it is observed. "Conciousness," he says, "is all we have left with which to create the universe."

◄ The nearest terrestrial equivalents to coral reef communities are tropical rain forests like Kalimantan, the Indonesian term for the island of Borneo. The Gaia hypothesis states that these intricate webs of chemical and biological cycles are part of a planetwide system that functions to perpetuate itself.

◄◄This coastal California river is an integral part of the great chemical cycles that link ocean, land and atmosphere. The sequoia trees that line its banks are among the largest and oldest living things on Earth.

▲ Winding among "peaks like a jasper forest," the Li Jiang carries dissolved carbon dioxide back to the sea.

▶ The limestone hills of Guilin, in the Guangxi Zhuang autonomous region of southern China, bear witness to both the creative and destructive powers of water.

▶▶ Guilin, which take its names from the flowering acacia trees that scent the city in autumn, was founded more than 2,000 years ago. For more than a thousand years its inhabitants have carved poems and images in the limestone cliffs of the Peak of Solitary Beauty and other famous hills.

The landscape of Southern China's Guilin region has inspired artists and poets for centuries. Flowering trees make hazy reflections in the misty Li Jiang River. Conical peaks rise steeply from the banks. This is the fantastic landscape of Chinese paintings and poetry. "The river is like a lady's belt, the mountains like an azure jade hair-stick," wrote Han Yu a thousand years ago.

Six hundred miles away at Shilin is another bizarre landscape. Known as karst by geologists and "melting rock" by the Chinese, it is a deeply fissured plateau marked by sinkholes, caves, and underground rivers. These remarkable formations, both the Li Jiang landscape and the karst of Shilin — as well as karst regions in Yugoslavia, the Yucatan Peninsula and the Alps — are made of limestone, a rock which is one of the most common on Earth and which in more prosaic forms covers nearly a quarter of the surface of China. Besides its role in artistic inspiration, limestone is a source of humbler products ranging from fertilizer to cement.

A STORE OF CARBON DIOXIDE

Its most important function, however, takes place without human intervention. Laid down over millions of years by marine organisms, the sediments that form limestone comprise a vast

storehouse of carbon dioxide. The amount of carbon in sedimentary rocks is more than 600 times the total carbon content of the planet's water, air, and living cells. Its slow accumulation and release is part of the drama of plate tectonics.

Ocean sediments are carried down into the mantle at the subduction zones between plates. There they melt in the radioactive heat, producing magma and gases that build up in chambers beneath the crust. Periodically the heat and pressure cause magma and hot fluids to breach the surface in the form of smokers, travertines, and volcanos. Volcanic gas may have carbon dioxide concentrations of more than 10 percent. For example, carbon dioxide released in eruptions at Heimaey was concentrated enough to kill cats on the street and render some houses uninhabitable. It accounted for the only human death in the volcano's four-month siege — a burglar who suffocated in a gas-filled drugstore.

Other rock, like that around Guilin, is pushed upward by the shifting plates. Four hundred million years ago, southwestern China was a tropical sea, home to a vast array of life. Prominent among its population was coral, a prodigious producer of calcium carbonate. Although under assault by pollution and predators, coral reefs are still the largest biological constructions on Earth.

◄ Guangxi Province has little arable land, but is compensated by landscapes of eerie beauty.

▼ Guilin at sunset.

103

A JASPER FOREST

1.

2.

▲ A "jasper forest" is formed:
1. Four hundred million years ago a warm ocean covered the Guangxi region of China. The calcium carbonate deposited by marine organisms created limestone strata on the ocean floor.
2. Raised above sea level by the movements of plate tectonics, the limestone was exposed to rain.
3. The slightly acidic rain seeped into cracks in the limestone dissolving weak sections in the rock.
4. As erosion continued, the water traveled further beneath the surface, collecting in underground rivers and pools.
5. Erosion continued to sharpen the surviving peaks.
6. As surface water worked its way ever lower, old caves and rivercourses were exposed to view.

3.

4.

5.

6.

▲ Not all limestone looks dramatic. This quarry provides material for cement and other utilitarian functions.

CORAL REEFS: BIOLOGICAL WONDERS

Fossilized remains of earlier reefs have emerged as limestone highlands in parts of Australia, Texas, and other areas. Today, the Great Barrier Reef off the coast of Australia covers an area of 80,000 square miles, and every acre of the colony produces between 100 and 500 tons of calcium carbonate a year.

When tectonic forces pushed the skeletons above the sea, water went to work once more. In a report published during the Cultural Revolution, Chinese scientists used the thoughts of Chairman Mao "On Contradiction" to describe the relationship of karst and water: "The interdependence of the contradictory aspects present in all things and the struggle between these aspects determine the life of all things and push their development forward." The dialectic of karst is based on the fact that it is both created and destroyed by water. The calcium carbonate of limestone is nearly insoluble in pure water, but if the water already contains some carbon dioxide, it can readily draw more from the rock. Rainwater contains enough carbon dioxide to begin the process.

IMPORTANCE OF LIMESTONE

The karst landscape of southern China was sculpted by the region's heavy rains. Small cracks in the limestone surface enlarged as water dissolved the rock. As the fissures deepened, the water coalesced into underground rivers beneath a jagged, barren plateau. Caves were hollowed out of the darkness where water pooled underground. When the dissolved carbon dioxide reached

◄▼ Limestone's role as a storehouse of carbon dioxide is demonstrated in a simple experiment. Rock from the quarry is placed in a bag and dissolved with a mixture of hydrochloric acid. When the bag is opened over a group of lighted candles, the carbon dioxide it contains promptly puts them out.

THE ANCESTRAL CELL

The search for our ancestors has long been a matter of rocks and bones. For the beachcomber, the elegant curves of a fossilized scallop shell in a sandstone cliff can embue an afternoon stroll with a deeper sense of time. For the paleontologist the shell may be interesting, but frustratingly new. Fossils of larger plants and animals go back about 600 million years, only a fraction of Earth's biological history.

Fossils of the microscopic life that represents our primal heritage are harder to find and harder to evaluate. A bacterium or other one-celled organism will create a fossil only under unusual circumstances. En masse, they have left a legacy in limestone and other rocks, but all but a few individual entities have disappeared without trace. Stromatolites are by far the most common record of microscopic marine life. Like an arrowhead or a stone knife, they are a production of a living thing but not the thing itself. Stromatolites do help guide the search, however. Without some visible trace, the search for microscopic fossils is a hunt for a diamond in a gravel pit.

The older the rock, the less likely it is that it will contain identifiable fossils. As eons go by, rocks — and the fossils they carry — are increasingly apt to be deformed, transformed chemically, or lost to the constant reshuffling of the Earth's crust. Certainly this is true of the Isua Formation in Greenland, the oldest known sedimentary rock. It has been heated to over 900 degrees Fahrenheit under high pressure, a metamorphosis sufficient to obliterate any record of early life.

Still, advancements in micropaleontology have opened up the Precambrian world of one-celled organisms and engendered new respect for their contributions. A very few fossils have been found in rocks more than 2.5 billion years old. The oldest incontrovertible evidence is from the Warrawoona Group of sediments near North Pole, Australia. The bacteria-like organisms preserved in the rock are some 3.5 billion years old. They can only whet the scientific appetite for more ancient examples, though, for even they are not the earliest forms of life.

"They already had a complex organization and were already of many different types," notes Dr. Malcolm Walter of the Australian Bureau of Mineral Resources. "So the origin of life must have been quite a lot more than 3.5 billion years ago."

While paleontologists push the study of life back into ancient rock, molecular biologists use computer programs to trace a path to life's beginnings. They are looking for even smaller messengers from the past — the DNA and RNA of ancestral cells.

▶ Dr. Malcolm Walter of Australia's Bureau of Mineral Resources examines some of the microfossils found at North Pole, Western Australia.

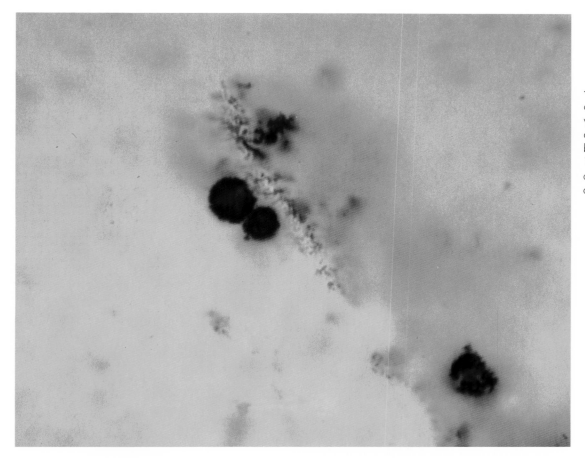

◄ The oldest known fossilized organism, dated at 3.5 billion years, is a collection of cyanobacteria. The two linked black spheres, shown here at 1,000 times magnification, and daughter bacteria immediately after cell division.

Advances in laboratory genetics now permit scientists to examine the specific sequence of amino acids in DNA. This breakthrough has allowed geneticists to apply one of the standard tenets of evolutionary biology, that organisms sharing the same trait were probably evolved from a common ancestor. This principle has long been used in constructing family trees from bones. Two dissimilar-looking beasts may reveal their relationship in a common arrangement of teeth or toes. DNA sequencing makes it possible to compare traits that have left no physical record. By comparing the chains of amino acids in different strands of DNA, it should be possible to trace species back to the specific mutation that caused their paths to diverge.

Family relations between a number of organisms can also be established by comparing the gene sequences. The more areas of duplication, the closer the relationship. Computers make feasible the complicated mathematics involved in these comparisons. Using a program that also adjusts for different rates of evolution among different species, James A. Lake, a molecular biologist at the University of California, announced in 1988 that he had found the universal ancestor of all life on Earth. His candidate is an Eocyte, a form of bacteria that survives today in boiling sulfur springs.

Paleomicrobiology is a contentious field, and Lake's theory has not won universal acceptance. Computer programs, however sophisticated, are only as good as the assumptions programmed into them, and some of Lake's colleagues find fault with his equations. Even more controversial was the announcement made in 1987 that anthropologists had traced a gene sequence back to a single human ancestor. Inevitably dubbed Eve, she is supposed to have lived in Africa some 200,000 years ago and to

have passed her genetic inheritance to everyone alive today.

The search for Eve involves mitochondrial DNA, the genes that power cell biology. Unlike nuclear DNA, which determines traits such as skin and hair color and is reshuffled with every instance of sexual reproduction, mitochondrial DNA is inherited intact from the mother. To follow its path back through the byways of human breeding, Rebecca Cann, a molecular biologist now at the University of Hawaii, collected placentas from 147 new mothers of various ancestry. She and her colleagues examined the DNA extracted from the placentas, comparing each link for similarities and discrepancies.

The differences in Cann's sample turned out to be minor. At the cellular level, a Siberian doctor and a Nigerian marketwoman may resemble each other more than they resemble their own neighbors. If Cann's calculations are correct, the whole variety of racial characteristics has evolved in the 200,000 years since Eve's daughters began passing on their genetic inheritance.

Some scientists, like Stephen Jay Gould, find this evidence of "biological brotherhood" profoundly exciting. Others, including some other DNA detectives, are not persuaded. Another study has found more racial differences in mitochondrial DNA and has placed the common mother in Asia rather than Africa.

Many paleontologists have more basic objections. They do not accept the proposition that only one of the human groups that lived 200,000 years ago managed to pass on its genes. Instead they place the universal ancestor back about a million years and allow for substantial genetic intermixing rather than the dominance of one group. It will take more than a computer printout to resolve the dispute. The quest for our beginnings may have just begun.

saturation point, it precipitated out drop by drop as pure calcium carbonate. Eventually the water, laden with 10 times the concentration of carbon dioxide in sea water, reaches the Li Jiang and returns to the China Sea.

What would happen if all the carbon dioxide locked up in limestone on the Earth today were suddenly released? The effects would be catastrophic. An Earth with 40 times as much carbon dioxide in its atmosphere would suffer a dramatic change in its climate, and might become more like Venus than the planet we know.

The oceans are a major key to the Earth's stability, not only in terms of carbon dioxide but also through other crucial relationships. Oceans are moderators of temperature, absorbing and redistributing huge quantities of solar heat. They are weather makers, interacting with the atmosphere to produce wind and rain.

▲ The karst formations around Shilin, in Yunnan Province, inspire rhetoric as improbable as their own vistas. According to the authors of *Karst in China* , "The broad masses of workers, peasants, and soldiers, together with the scientific and technical personnel, take class struggle as the key link, conscientiously study the theory of the dictatorship of the proletariat, keep firm to the Party's basic line, implement earnestly Chairman Mao's important instructions . . . and insist on combining boundless revolutionary enthusiasm with strict scientific attitude. Thus they have made new achievements in recognizing, utilizing, and transforming karsts."

▶ Water finds its level in a pool studded with stone towers.

▶▶An elephantine silhouette, like the notch below it, shows the path taken by water through the dissolving limestone.

OXYGEN CONTENT

Without life, the atmosphere would have no oxygen. Without oxygen, life on Earth would be limited to the inhabitants of volcanic vents and mudflats. Our need is constant and specific. Our cells cannot go more than a few minutes without the energy oxygen supplies, nor can they cope with drastic changes in concentration. Premature babies have been blinded by too much oxygen in incubators. Pilots and mountain climbers have died from ailments and errors of judgment brought on by too little.

Researchers have carefully mapped the boundaries of our tolerance, but still we know relatively little about the history of oxygen in the air. Unlike rocks, which may be preserved for eons, atmosphere is evanescent. Glacial ice has trapped air dating back 160,000 years, but that only amounts to a blink in the scale of geologic history.

One possible repository for ancient air is amber. The tree sap or resin that fossilizes into amber can function like a miniature tar pit. As it oozes from a break in the bark it may trap insects, small animals, and air bubbles. In 1987, Gary P. Landis of the U.S. Geological Survey and Robert A. Berner of Yale analyzed air from amber of different ages by crushing samples in a vacuum and running the trapped gases through a spectrometer. They found a surprising range of oxygen levels. The highest — about 30 percent — was found in a sample dated at 80 million years. A 40-million-year-old sample yielded modern levels, while one from 25 million years ago was under 20 percent oxygen.

Although the reports were preliminary, they correlated well with predictions made by experts such as Heinrich Holland of Harvard based on the cycles of weathering, which lock up oxygen by oxidizing iron compounds, and the burial of organic matter in the ocean, which correlates with rising levels. Holland expected higher oxygen levels about 80 million years ago during the Cretaceous period, the time of the break-up of Pangaea, when there were many shallow seas and a corresponding reduction in dry land. That meant more area for the burial of organic matter in marine sediment and less weathering from surface erosion.

The amber samples may not be valid. Experiments by other chemists have indicated that the resin is too absorbent to keep an air sample pristine for millions of years. However, the prediction of high oxygen levels during the Cretaceous period gets indirect support from the fossil record. Some paleontologists think that the massive size of dinosaurs and other Cretaceous creatures was made possible by an oxygen-enriched atmosphere which made cell biology more efficient.

It is also possible that the same levels that helped the giant beasts thrive contributed to their demise. James Lovelock, originator of the Gaia hypothesis, says that levels of even 25 percent would turn the Earth into a giant firetrap, where any spark could lead to a cataclysm. Even if he is wrong in his specific percentage, there is no doubt that oxygen-enriched air makes fires burn more fiercely. This may help explain the evidence of wildfires found at the K-T boundary. Soot particles found in the boundary layer have led researchers to postulate a continent-wide blaze set off by a meteorite impact. If oxygen levels really were higher 65 million years ago, this scenario could gain credence.

▶ These cliffs at the shore of Great Slave Lake are made of fossilized stromatolites. The bacteria that made them thrived in incredible numbers in the early oceans, and the oxygen they released transformed the world.

Atmospheric oxygen may also rise and fall according to the sea level.

Although the total amount of water has remained constant for eons, its distribution over the planet has changed. As plate movements rearrange the continents, oceans become deeper or shallower, with longer or shorter coastlines. When coastal length increases, more shallow water allows more photosynthesis in the ocean, more production of oxygen, and more burial of carbon in ocean sediments. This was the situation during the late Cretaceous period 110 million to 85 million years ago. Smaller, deeper seas of the sort we have now

▲ "Melting rock" in a cave at Shilin marks a balance point in the carbon cycle as water becomes saturated with carbon dioxide. In the caves carved by underground rivers, calcium carbonate dissolved from the rocks above precipitates back out into stalactites and stalagmites.

◀ A research team explores an underground river near Shilin. Its water contains ten times the normal percentage of carbon dioxide.

◄ Continued erosion has exposed the former course of an underground river.

◄◄ The hills and rivers of Guilin are only part of an immensely complex cycle involving the deposition and release of carbon dioxide.

may correspond with lower oxygen levels. But these correlations are far from straightforward. Lower sea levels also mean that more land is exposed, which means more erosion, which, in turn, means that more phosphates will be carried into the sea. Increases in phosphates are linked to higher oxygen production and carbon dioxide consumption by oceanic plants.

It is difficult to evaluate the effect of a change such as the increase in ocean phosphates. The variety of Earth's chemical cycles makes it unlikely that any one factor will cause a drastic alteration in oxygen or carbon dioxide levels. That same reassuring complexity, however, may make us unaware of a dangerous trend until it is too late to reverse it.

◀ The clouds rising over the island of Tarawa are part of a weather system that affects climate all over the Earth.

▶▶A cumulonimbus cloud rises above Tarawa. The flat base marks the onset of condensation as the warm, humid air reaches a colder layer of the troposphere.

PATTERNS IN THE AIR

FROM the Moon, which has no atmosphere, the sky looks black. From the plains of Mars, where the thin air is mostly carbon dioxide, it appears orange-red. The composition of Earth's atmosphere gives its sky and its oceans an azure hue. For 300 miles until it dilutes into the scattered molecules of space, the atmosphere encloses Earth in a life-preserving cocoon. Hurricanes and hailstorms are part of the price we pay for this protection.

Atmosphere is a blanket term for a multilayered composition of gases and suspended particles. Seen through the window of the space shuttle, the layers show as a succession of blue bands against a darkening sky. The shuttle's journey through the bands at 1,800 miles per hour takes 10 minutes.

The first layer is the troposphere, home of winds and weather. Only about 10 miles thick at the Equator and half that at the poles, it contains about 80 percent of the mass of the atmosphere. A cargo of solid particles including dust — both earthly and meteoric — seeds, sea salt, smoke, and radioactive detritus from nuclear explosions is suspended among the water vapor and other gases. Temperatures are warmest near the surface. By the time the shuttle reaches its outer boundary, the tropopause, the thermometer reads below –100 degrees Fahrenheit.

The next layer, the stratosphere, extends to about 30 miles above the ground. A thin band of stratospheric ozone absorbs ultraviolet light, sending the stratospheric temperature back above freezing and shielding the Earth below from dangerous radiation.

The mesosphere, between 30 and 50 miles above Earth, is the region where incoming meteorites start to burn. It is also the upper limit for water vapor. These last traces of evaporation refreeze as they rise from the stratosphere. Occasionally enough vapor collects to form noctilucent clouds that glow eerily in the night sky. The mesosphere is the coldest layer, with temperatures of –250 degrees Fahrenheit along its upper boundary.

▼ Daily rains bring lush growth to the islands of Oceania, but most of the soil is thin and unsuitable for agriculture. Villagers supplement their marine harvest with coconut and breadfruit.

▲ Seen from the space shuttle, the atmosphere is a fragile, glowing blanket, shielding Earth from the unforgiving blackness of space.

Two minutes into its flight, the shuttle leaves what can really be considered as air. Less than 0.00001 percent of the atmosphere is in the thermosphere, which is five times as deep as the first three layers combined. The thermosphere and the exosphere, where gravity loses its hold and the last traces of atmosphere merge into space, are the sites of intense bombardment from ultraviolet radiation. The resulting electromagnetic bands have the prosaic names of D, E, F1, and F2, but they create the magical tapestries of the aurora borealis and aurora australis. They also are important in radio communication.

All the planets — and some of the moons — in our solar system have at least a trace of an atmosphere. Their initial characteristics were determined by the planets' size and temperature, the factors that determine which gases are retained as gas. Once formed, atmospheres are not static but evolve along with the planets. Earth's air is nearly 21 percent free oxygen, and its surface temperature stays generally within that narrow band where water can exist in liquid form. But it was not always so, and may not necessarily remain so forever. Mars, now a freezing desert, may once have had liquid water too.

LAYERS OF AIR

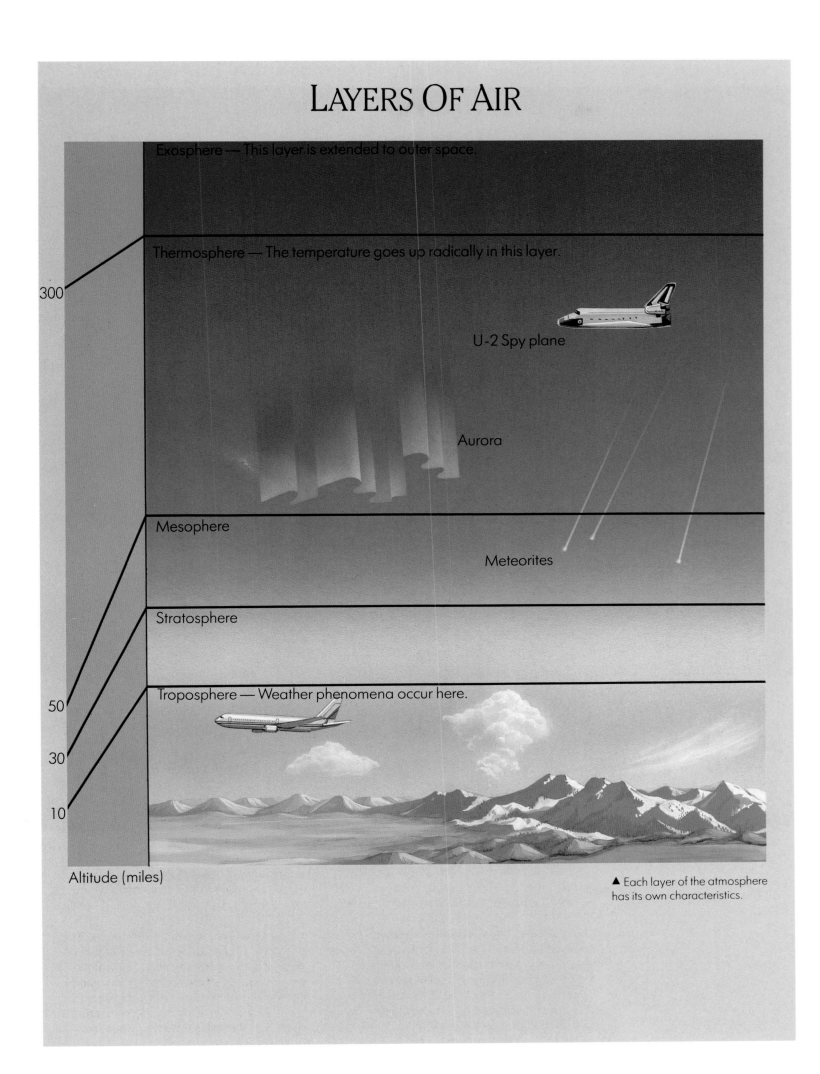

Exosphere — This layer is extended to outer space.

Thermosphere — The temperature goes up radically in this layer.

U-2 Spy plane

Aurora

Mesophere

Meteorites

Stratosphere

Troposphere — Weather phenomena occur here.

300

50

30

10

Altitude (miles)

▲ Each layer of the atmosphere has its own characteristics.

121

▲ The Colorado River is still cutting deeper, as evidenced by the sediments that color it.

ROCKS AND THE ATMOSPHERE

Rock formations on Earth can be read as a sort of yearbook of atmospheric evolution. The walls of the Grand Canyon have been cut over the last 15 million to 20 million years by the muddy waters of the Colorado River. The layers of rock exposed by the river are much older than that. Rocks at the top of the canyon walls are roughly 200 million years old, while those at the bottom of the gorge were formed closer to two billion years ago. Even now the river continues to cut deeper.

The lowest rocks were created around the transition point to an aerobic atmosphere. One of the markers of the change is the presence of banded iron formations, sedimentary rocks of iron oxide and silica. They indicate an ocean rich in dissolved ferrous iron, which can only exist in the absence of oxygen. It took as much as a billion years of photosynthesis to create the oxygen that transformed ferrous iron to ferric iron, replacing banded iron formations with "red beds," such as

the oxidized iron deposits of Minnesota's Mesabi Range and the Hamersley region of Western Australia.

The world in which the red beds were created was poised on the border of a great change. In the ocean the presence of oxygen was fueling tremendous evolutionary experimentation, but the land was still barren. Clouds cast their shadows, glaciers advanced and retreated, rivers dug channels in the rock. But no grass waved in the wind. No roots soaked up rain. There were no organic leavings to turn sand into soil.

ULTRAVIOLET RADIATION

Before the build-up of atmospheric oxygen, ultraviolet radiation reached the Earth unimpeded. Water provided a shield against its destructive power, protecting the vulnerable genetic material in cells. Some photosynthetic organisms stayed near the surface to harvest sunlight and developed a substance known as sporopollenin, which filters

▲ Fossilized bark of a tree-sized *Lepidodendron*, a plant grouping which has survived from the Devonian to the present day in the inconspicuous form of club moss.

◄ The delicate whorls of *Equisetum*, from a Devonian fossil, are recognizable today as horsetails, the only living genus of a major Devonian plant division.

TREES OF HISTORY

THE geological record of Earth's early atmosphere is equivocal. The conditions that contributed to banded iron formations, for example, are not adequately understood. And in the case of red beds, red rocks can be deceiving. Some of the ruddiest cliffs of the Grand Canyon are actually composed of grayish limestone, with only a surface stain of oxidized minerals.

But some of the stone legacies are much easier to read, though hard at first to believe. The petrified trees found throughout the world, and seldom more spectacularly than in northeastern Arizona, are exactly what they appear to be, the remains of a forest preserved from another time.

Some 175 million years ago, Arizona was part of a subtropical flood plain. Instead of the cacti and scrubby grasses that grow in the Painted Desert today, the area was dominated by giant ferns, mosses, and other moisture-loving plants. Amphibians and crocodile-like reptiles, some of them more than 20 feet long, dozed in the swampy shallows.

On slightly higher ground grew trees, mostly primitive conifers, ranging to over 100 feet in height. Floods were frequent, and trees were often uprooted and washed downstream, to come to rest in giant logjams. Over millions of years, more than 400 feet of sandy mud and volcanic sediments were deposited on the flood plain, interspersed from top to bottom with downed trees.

Sometime during the Jurassic period that ended 136 million years ago, the floodplain sank, much as the Great Rift Valley is sinking today. The high ground remained for a while as scattered islands, and then disappeared under a sea that stretched all the way to present-day Delaware. The sea remained for 100 million years, covering the mud, sand, and trees with thousands of feet of sedimentary rock.

Eventually the forces that raised the Sierra Nevada and Rocky Mountains, also pushed up the Colorado Plateau. Marine rock was exposed to the winds and sudden rains of an arid region. As it eroded away, the top layers of trees, now turned to stone, surfaced once again.

▲ Less resilient than the original wood, petrified trees are often found broken, usually into more-or-less equal segments. The favored explanation is that vibrations from earthquakes in the area have put too much stress on the brittle silica formations.

▶ Log segments that once lay beneath the sea now top rock formations in the desert.

▲ Petrified Forest National Park. Far from any modern grove, fossilized trees are scattered across the Painted Desert of northeastern Arizona. A Navajo legend describes them as the legs of a giant from ancient times.

◀ As erosion continues, more trees are uncovered. Denser rock formations also are revealed when the softer shales above are stripped away.

▲ Before life moved landward from the oceans, the increasing oxygen levels created by photosynthesis began to oxidize the rocks, reddening Earth's surface until it may have resembled Mars. Without any vegetation to decompose, there was little soil, and what meager sediments may have formed through chemical processes and weathering had little chance to build up.

▶ The efficiencies of oxygen-powered metabolism allowed marine plants to grow larger and more diversified. Some groups, such as bryophytes, mosses, and liverworts, developed the ability to go dormant during dry periods without destroying their delicate reproductive bodies. Thus equipped they were able to begin colonizing shorelines and river deltas.

out ultraviolet radiation and protects against water loss. This innovation helped pave the way for a terrestrial transition, but the oxygen created by the early photosynthesizers was even more important.

Under ultraviolet bombardment in the stratosphere, oxygen molecules split and then recombined into an unstable grouping of three atoms. This gas, called ozone, formed a thin, ever-renewing layer that absorbed large quantities of ultraviolet radiation. By 1.4 billion to 1.2 billion years ago, the ozone layer was effective enough to permit life on land. Algae and bacteria probably made the first move, colonizing pockets of moisture. They were followed by amphibious species that lay dormant during dry periods. As the first terrestrial organisms flourished and died, their decomposition created food and soil for the next wave of colonists.

The earliest known example of a vascular plant is the unprepossessing *Cooksonia*, which appears in fossils from the late Silurian period 400 million years ago. Their distribution pattern provides some evidence for the existence of the supercontinent Pangaea. By the Devonian period about 40 million years later, the main groupings of higher plants had appeared. Seed producers, with their reproductive

material encased in a protective coating, had an unprecedented ability to withstand changes in climate and soon became the dominant landform.

AMPHIBIOUS INVERTEBRATES

Sometime during the Devonian, the first amphibious invertebrates crept from the water's edge. Equipped with a hard external skeleton and multiple legs, these ancestors of the millipede found ample forage in the mats of mosses and

primitive vascular plants. Some grew to be six feet long.

Despite the ozone shield, genetic material was still vulnerable to damage from solar radiation. Today, some frog species have a curious characteristic that may shed light on how the early land colonizers dealt with levels of ultraviolet radiation higher than we know today. The eggs of the leopard frog are white on one side and black on the other. The white part contains important genetic material. The black part contains a barrier

to ultraviolet light. When exposed to light, the egg rotates to present its dark surface, blocking harmful radiation and collecting heat that helps with incubation.

WEATHER

Earth receives tremendous quantities of solar energy daily. About 20 percent is absorbed by the atmosphere, and another 30 percent is reflected off clouds, snow, and other light-colored surfaces. The remaining 50 percent is absorbed with varying degrees of efficiency by the planet's patchwork of land and sea, field and forest. This unequal distribution of heat is inherently unstable. The process by which this energy seeks its equilibrium is what we call weather.

Although weather often defies prediction even in an era of satellite photos and computerized maps, its general patterns have been observed for centuries. Salish Indian families in the Pacific Northwest still use the hot wind funneling down the Fraser River canyon to dry their salmon catches. In the winter when that same wind bred blizzards, their great-grandparents used to say that the North Wind people were hurrying home to

▲ The advent of vascular plants allowed flora to move inland. By the Devonian period, 380 million years ago, waterways were lined with a thick layer of greenery. Taller species took root in sediments laid down by their Silurian predecessors.

◄ By the Silurian period 400 million years ago, plants had evolved enough structural framework to stand without the support of water, and enough ozone had been created in the stratosphere to protect organisms from ultraviolet radiation.

◀ A Devonian forest. The dominant plants were club mosses, horsetails and true ferns. Because all these types lack seeds, they require a moist environment for their sexual phase, in which the sperm needs at least a little water to reach the egg. Therefore the uplands were bare. Seed plants have more versatility, since they can be fertilized without standing in water.

▲ Artist's rendering of *Cooksonia*, a widely distributed mosslike genus that is the earliest vascular plant identified in North America. Although small and spindly, it contained the cell bundles that allowed it to stand upright in air. Examples of *Rhynia*, another vascular genus that dates from the same period, have been found in Australia.

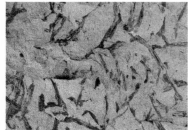

▲ A magnified photo of fragments of *Cooksonia* fossils from Silurian sediments in Wales. The rounded spore pods were actually the size of pinheads.

RAINFALL PROJECTIONS

▲ Rainfall models are made and examined at Japan's most famous meteorological observatory, in Tsukuba, Ibaragi Prefecture.

► Average daily rainfall is shown on this map, expressed in milimeters. The palest blue is 0 to 2 mm per day, with each gradation indicating another 2mm.

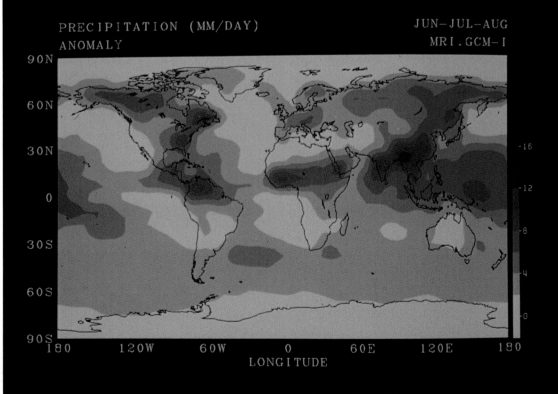

PRECIPITATION (MM/DAY)   JUN−JUL−AUG
ANOMALY   MRI.GCM−I

◀ Dr. Tatsushi Tokioka, right, works with rainfall projections.

◀ Rain forests help moderate rainfall elsewhere on the globe. The red areas on this map indicate areas of decreased rainfall according to a computer model, if the rain forests were cut. Darker blue areas denote an increase.

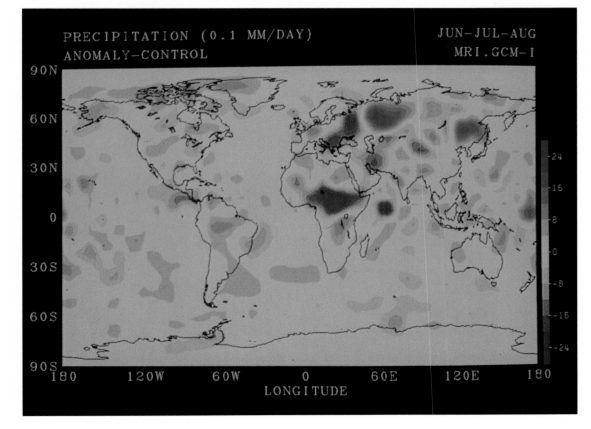

PRECIPITATION (0.1 MM/DAY)   JUN-JUL-AUG
ANOMALY-CONTROL   MRI.GCM-I

90N
60N
30N
0
30S
60S
90S
180   120W   60W   0   60E   120E   180
LONGITUDE

24
16
8
0
-8
-16
-24

WIND PATTERNS

rescue a child from a negligent babysitter. Aristotle, only slightly less fanciful, called wind the dry sighs of the breathing Earth.

In the tropics, where the trade winds begin, the creation of wind is visible. Located in the Pacific Ocean northeast of Australia, the island nation of Kiribati is bisected by both the Equator and the international dateline. Its capital is Tarawa. Here along the Equator is where the air currents that drive much of the planet's weather originate. The islanders still depend on the sea as they have for centuries, pursuing fish in sail-powered outriggers. Children learn the ways of the wind with toy boats, preparing for the day they will venture into the open ocean.

Once away from the islands, the sky impresses as much as the sea. Even on the sunniest days, clouds put on a dramatic show around Tarawa, changing hour by hour. Each morning the tropical sun evaporates large quantities of ocean water. When this warm, moist air rises to higher elevations it cools and its moisture condenses into the elaborate cumulus battlements of the lower atmosphere. Carried upward by the same heat that created them, they may rise into the lower stratosphere, where the rapid drop in temperature turns those benign, fluffy puffs into towering thunderheads. These cumulonimbus clouds carry an unstable load of supercooled water and ice, which they release as thunderstorms and hail.

CLOUDS

As the clouds rise around Tarawa, they create a low-pressure area. Air moves in from the northeast to fill it, a phenomenon that was exploited long before it was understood. Tarawa marks the southern extreme of the trade winds that sent Columbus and Magellan across the ocean. The trades are part of a worldwide system by which disparities in the Earth's heat are mediated by the atmosphere. Their corollaries are the still air of the equatorial doldrums and the horse latitudes.

DOLDRUMS

In the doldrums, the rising column of heated air is so steady that it blocks incoming currents. Sailing ships may drift for weeks in a sultry calm broken only by thunderstorms that release rain from a saturated sky. The air that rises in the

▶ From the cloudless Sahara to the hidden ice sheets of Antarctica, all processes on Earth are connected through the movements of the atmosphere.

◄ The global cloud cover is shown here in satellite images taken over a week in June 1987. The clouds over Tarawa are part of the band encircling the Equator. Tropical marine air rises, produces clouds, and releases rain. After it has jettisoned its moisture, the air travels along the upper level of the troposphere until it comes back down around latitude 30 degrees in both hemispheres. These latitudes are the clear bands above and below the Equator on this globe.

◄ Marine temperature readings show the inequalities in Earth's receipt of solar energy. The warmest water is red; the coldest is dark blue. Without the currents of air and ocean, temperature disparities would be even greater.

◄ The same weather patterns that govern life at Tarawa create rainforests such as the Kalimantan on the island of Borneo. Water is the key to the exuberance of the rain forest, and Kalimantan receives between 100 and 200 inches of rain a year.

▲ The interplay of air and water around the islands of Kiribati affects weather thousands of miles away.

▶ Evaporated moisture hangs over the ocean in ranks of cumulus clouds.

doldrums descends in the horse latitudes, stripped of its moisture by its journey through the stratosphere. These regions of low humidity and little wind were named by desperate sailors who jettisoned their cargo of horses to save water.

UPDRAFTS AND JETSTREAMS

From the time when these alternating belts of wind and calm were first recognized, people have sought to understand them. Edmund Halley, better known for the eponymous comet, came up with part of the answer by 1686 when he identified the updrafts over tropical seas as the source of the trade winds.

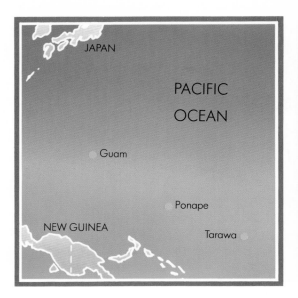

JAPAN

PACIFIC

OCEAN

Guam

Ponape

NEW GUINEA

Tarawa

This circular movement is the most powerful of the Earth's wind systems, spawning not only the trade winds but also the high-altitude jet streams. Racing at up to 200 miles per hour, these powerful currents occur where changes in temperature and air pressure are most abrupt, at about 30,000 to 45,000 feet above the boundaries between hemispheric temperature zones. First confirmed by World War II pilots, the jet streams now are used to save time and fuel on commercial flights.

EARTH'S ROTATION

The direction of the jet streams and the trade winds is determined by the rotation of the Earth. George Hadley first outlined the effect of rotation on air movement in 1735. Further refinements were announced in 1835 by Gaspard Gustave de Coriolis, who worked out a series of equations

▲ Tarawa at sunrise

◄ Sunset colors clouds that are still rising from the heat of the afternoon.

◄◄By the late afternoon, the sky near Tarawa has a complex geography of clouds, ranging from the cumulonimbus that is delivering a rain squall at the photo's center, to the altocumulus glowing white at center right.

THE AURORA

THE aurora are among the most awesome phenomena visible on Earth. A typical display in the polar regions commences with a greenish-white light arcing from horizon to horizon across the apex of the sky. Luminescent beads of light run along the arc at tremendous speed, while their shimmering pathway oscillates like a plucked guitar string.

Change is the only constant in this natural light show. Within seconds the configuration and colors of the aurora can change drastically, either subsiding to a dim northern glow or growing into massive curtains of light that seem to sway on an unseen wind. White tinged with greens and blues are the most common colors, but occasionally there are eerie red aurora.

Although predominantly a polar phenomenon, the aurora borealis has been observed as far south as the Equator. When, during the last two centuries, particularly strong aurora have appeared over populated areas, they have created havoc. In 1859 one of the strongest aurora on record knocked out telegraph communication in many areas of Europe by producing a continuous attraction that overwhelmed the pulses of electromagnets. In the United States it was possible to send messages from Boston to Portland, Maine, without a battery,

using only the power that accompanied the aurora.

The first person to realize the source of the aurora's power was Richard Carrington, an English physicist and astronomer. On September 1, 1859, Carrington observed a huge solar flare on the surface of the Sun. When intense aurora covered most of Europe two nights later, he immediately suspected the Sun was the cause. Carrington's insight established that it was not ice, nor radium, nor torchlight on angels' wings, but mighty explosions on the Sun that lit the northern lights.

The link between the Sun's surface and Earth's sky is the "solar wind," a continuous flow of charged subatomic particles from the Sun that streams around the Earth. When the solar wind passes through Earth's geomagnetic field in the upper atmosphere, it generates electrical energy that drives ionized particles into the denser gases below. The greenish-white color of most aurora is no mystery; it is the color given off by oxygen when ionized.

The electromagnetic bands that generate the aurora are in the thermosphere about 65 miles above sea level. Viewed from space the aurora borealis and australis appear as halos around their respective poles. They may expand in diameter and

▶ The swirling traceries of the aurora are caused by "solar wind" passing through the electromagnetic belts of the thermosphere. Simultaneous observations taken in Iceland and Antarctica have revealed mirror-image auroras in the northern and southern hemispheres.

▶▶ Greenish white, the color of ionized oxygen, is the most common manifestation of the aurora.

distance from the poles, but they are almost always present. On the ground beneath them, a variety of strange side effects plague human equipment.

The electrical currents emanating from the aurora can accelerate rusting. As a result, oil and gas pipelines in high latitudes develop pinhole leaks. The Haines Pipeline in Alaska was rendered useless in a little more than a decade, and aurora is a concern over the Alaska Pipeline. The powers that produce the aurora also disturb the early warning defense systems of the United States and Soviet Union. Some advanced forms of radar become ineffective when the aurora dimples the reflective layers of the upper atmosphere.

In fact, we are vulnerable to the aurora as never before. Only since the development of large-scale electrical systems has the aurora been able to touch the lives of humans directly. The extent of this power was evident in February 1958, when a large part of northeastern Canada was blacked out by an auroral "superstorm" that overloaded utility circuits.

Human dependence on electrical devices has increased since then — especially on computers, which are particularly vulnerable to disruptions in power — but so far there have been no comparable storms. "Statistically, we are overdue for a superstorm," says Dave Speich of the U.S. Space Environment Service Center. "There are 20 auroral superstorms on record since 1880 and none since 1960."

The next time the aurora come in force, they may make an impression on human culture unlike anything we have seen before.

▲ A scientist examines peat in an Okefenokee swamp.

▶ Not every cypress swamp becomes a coal bed. In order for coal to form, the swamp must periodically be covered by the sea. Layers of marine sandstone and shale lie over each coal bed. The process is underway now in parts of the Mississippi Delta.

the polar regions, the Earth would be much hotter at the Equator and much colder at the poles than it is today.

The climatic extremes of Earth can be seen as both causes and effects of the atmosphere in motion. The presence of life, with its own cycles of water and gases, is a major factor in the equation. At Kalimantan, Indonesia, a vast rain forest luxuriates in the same conditions that produce the winds of Tarawa. Like ocean water, the morning fogs and dews of Kalimantan evaporate as the Sun intensifies and return in afternoon downpours that amount to 100 to 200 inches of rain a year. Prolific vegetation protects the thin soil against erosion, contributing to an ecosystem that is exceptionally successful at capturing and reusing water. The

◄ South of these green Algerian foothills, the Sahara begins.

◄◄ Moisture from the Mediterranean gives northern Algeria a climate similar to the south of France. In the background, the Atlas Mountains strip the moisture from the coastal winds, and mark the boundary of the arid regions to the south.

result is a climate of impressive stability, where temperature, humidity, and rainfall hardly vary throughout the year.

DESERT WINDS

Enacted over a different landscape in the same latitude, the interplay of wind, sun, and water can produce desolation. When the Sun shines on sand instead of trees, there is little water to extract and little cloud cover to moderate the intensity of radiation. Average temperatures in the Sahara desert may not be much higher than in the rain forest, but the variation is extreme, sometimes as much as 100 degrees Fahrenheit in one day.

Deserts are found on every continent between 15 and 30 degrees latitude. The largest is the Sahara of North Africa. *Sahara* means deserts in Arabic and

▶ The climatic logic of the Hadley Cells helps impose the desert's destiny. The scorching winds of the Sahara arrive back at the Equatorial latitudes ready to pick up another load of moisture and begin the cycle again.

does actually comprise a collection of deserts with differing characteristics. It is the home of winds like the *khamsin, harmattan,* and legendary *sirocco,* which sometimes cross the Mediterranean to bedevil southern Europe. In the desert, winds bring drought instead of rain, creating connoisseurs of thirst. The people of the Sahara distinguish at least eight different degrees of thirst, culminating in *al-Hayam,* "a vehement thirst" and also a synonym for passionate love.

THE HADLEY CELL

The ferocious winds of the Sahara are part of the same cycle that creates the lush vistas of Tarawa and Kalimantan. The system was described by George Hadley nearly 250 years ago and consequently has become known as the Hadley Cell. Air along the Equator just south of the Sahara rises in the tropical heat. As it reaches the upper troposphere it cools, releases the moisture it has accumulated, and begins to sink back toward Earth at about 30 degrees latitude north and south. The low pressure at the Equator draws the air back to be heated and rise again.

Air sinking over the desert finds no moisture, and the balmy trade winds of ocean regions become desiccating scourges like the *simoom,* or "poison wind" of the Arabian desert. The contrast is apparent in North Africa. Algeria's northern coastline, watered by moisture carried from the Mediterranean, is green and fertile. South of the Atlas Mountains, the Algerian landscape changes drastically. Within minutes of sunrise in the Sahara, the night chill — sometimes as low as 20 degrees Fahrenheit — is vanquished by the Sun. Temperatures may rise to over 130 degrees Fahrenheit in the shade, but since shade is a hypothetical concept in much of the Sahara, the actual heat is much greater. Ground surface temperatures have been recorded as high as 183 degrees Fahrenheit.

As the heated air rises, the *harmattan* springs to life, sweeping huge quantities of sand and dust into the air. A Saharan dust storm can carry more than 100 million tons of debris, depositing some of the smallest particles as far away as South America and Central Europe. The "blood snow" that sometimes falls on Paris is colored by Saharan dust.

▶▶ Blood snow falls on Paris. In 1969, enough dust reached England to tint cars red overnight.

▶ A funnel on a Parisian rooftop is part of the monitoring system for desert dust.

▲ Wind and gravity are the sculptors of the great dunes of the Sahara. The ripples on the advancing slip face of the dune are formed by sliding grains of sand that have reached their angle of repose.

►►The sediments that once covered the volcanic rock of the Hoggar Massif have long ago been washed away by the once-frequent rains, leaving these stone battlements.

▼ Windblown sand is gathered into great drifting seas, called *ergs.*

SAND DUNES

The sand that remains in the desert falls into the lee behind rocks, shrubs, or other windbreaks and is sculpted into dunes. Without an anchor to the rocks beneath, dunes are inherently unstable. As grains slide down their steep leeward slopes, the whole formation gradually shifts. Some types can travel across the rock at rates of over 15 miles per year. In some areas the dunes coalesce to form seas of sand called *ergs.* They can reach vast proportions; one *erg* in Libya is as large as France.

The dunes of the Sahara began as rock. One source is the Hoggar Plateau of southern Algeria, located in the "dead heart" of the desert where rainfall averages less than half an inch per year. On the plateau basalt columns rise in battlements above a rock-strewn plain. The basalt is a legacy of

▶ Although sand particles are moved and shaped by the wind, the main agent that turns rock to sand is water.

ancient volcanic eruptions. Over the millennia, weathering and erosion have turned its covering of sandstone and granite to sand and dust. Unlikely as it seems today, much of the erosion was accomplished by water. Rivers flowed in the Sahara as recently as 10,000 years ago. Oak and cedar once grew in the desert, and the grainfields of North Africa were among the prizes of the Roman Empire. The oases used by camel caravans and nomadic tribes today are supplied by the underground water remaining from this wetter time.

A VICTIM OF THE DESERT

The Sahara blooms and wilts according to the same rhythms that govern the glaciers of higher latitudes. In its interior, climatic forces are so extreme as to mock attempts at human intervention. The forces of desertification are more complex in the Sahelian region along its southern range. In the south Saharan nation of Mauritania, where drought has lowered the water level so much that only the very deepest wells still produce, hundreds of thousands of people and 80 percent of the nation's livestock have died since 1968. The 700-mile "Road of Hope," constructed to bring the produce from desert settlements to the more populated coast, has become a highway of desperation for people with the wind at their backs.

The refugees converge on the coastal city of Nouakchott, whose name comes from one of the shallow Saharan lakes, called *chotts,* that are being lost to the desert. The capital of a country the size of France and Spain combined, Nouakchott has grown from 12,000 people in 1964 to more than 350,000 today. Sand drifts in the streets, along with

▲ These Mauritanian herdsmen have traveled for five hours to reach this well, the closest one to their drought-stricken homes.

▶ Vegetation and even level patches of stone alter wind patterns enough to create distinctive patterns of dunes. Symmetrical, crescent-shaped barchan dunes develop in areas where the supply of sand is limited, and the wind generally constant in one direction. Where sand is more abundant, long transverse dunes lie perpendicular to the prevailing winds. Seif dunes, stretching over many miles and reaching heights of a hundred yards, run parallel to the wind.

mounds of rubbish as the *harmattan* scours through the refugee encampments and on to the Atlantic.

Many climatologists feel that the drought in the Sahel is feeding upon itself. As vegetation is stripped by overgrazing and overcutting of firewood, the soil's surface dries out, creating less potential for rain and reflecting more heat. No longer held by an active root system, soil succumbs to the pressure of wind and sets sail.

▲ Sheltered between dunes and a stony outcrop, Taghit, Algeria, is one of the oases of the northern Sahara. The thick-walled courtyards offer some relief from the heat.

◀ Dust fills the air in a refugee settlement at Nouakchott, Mauritania. Many families have made dwellings out of shipping crates.

▲ The Road of Hope is being engulfed by sand. It is cleared by bulldozers, and when they are not available, by hand shovels.

▶ Nouakchott, built when Mauritania gained independence from France in 1960, is one of the most carefully planned cities in Africa. But the arrival of hundreds of thousands of refugees, and the encroaching desert that forced their relocation, has surrounded the central city with a desolate community of tents and shipping crates.

The rare rains that once replenished wells now race by in destructive flash floods. The process of desertification, once established, is difficult to reverse.

DESERTIFICATION

The Sahel is by no means the only region where human activity is exacerbating climatic trends. In the equatorial rain forest an estimated 77,000 square miles a year are falling to the chainsaws. This massive deforestation may have a significant effect on the global climate. Like Kalimantan, other tropical forests create much of their own rain through transpiration and evaporation. When the trees are cut the source of water goes with them, and annual rainfall may diminish to one third of its former level. Nor are the effects of deforestation limited to the areas cut. As the drier air over the former forests cycles through the atmosphere, it brings less rain to temperate regions too.

Computer projections predict lower rainfall in Eurasia and North America if deforestation along the Equator continues at its present pace. Compounding the problems caused by overcutting is the burning that accompanies it. Slash-and-burn agriculture is a high-speed variation on the carbon cycle exemplified by the limestone along the Li Jiang. Trees lock carbon into their cell structure. Burning releases it, using up oxygen in the process

CHANGING CLIMATE

Thousands of species that once thrived on Earth are gone today, victims of a change in temperature or environment. Whether by instinct or conscious design, the survivors have found ways to adapt or manipulate their immediate environments for their own benefit.

For millions of years, designated worker bees have fanned the entrance to their hives in hot weather. A hundred thousand years ago, near the encroaching glaciers of the last ice age, human communities assigned someone to keep the fire going in the family cave. The bees have kept to the same routines, but human efforts to alter their climate have become considerably more ambitious since then.

Asian farmers learned long ago to sprinkle their snowy fields with dirt or ash to hasten spring planting. But the same snow that is an impediment on the field can be a resource a bit farther up the valley. Summer melting of glaciers can provide fresh water in the season when it is most needed. Japanese scientists have been working since the 1960s to create a glacier out of a snowfield in the mountains south of Tateyama. If they can divert enough snow onto the field in winter and keep enough of it from melting away in summer, they may someday have a new source of water and cool breezes for Chiba Prefecture. Meanwhile, lacking any prospects for creating glaciers of their own, some Arab countries have proposed towing Antarctic icebergs through the Persian Gulf and docking them at desert ports.

There are many less ambitious ways to ameliorate or adapt to local extremes in climate. Every Midwestern farmer knows the value of windbreaks and shade trees, and even the backyard gardener soon learns to use the corn patch to shade the summer lettuce. A knowledge of microclimates is one of the requisites of good farming. When small successes ignite the human ambition for bigger things, however, the outcome is much less certain. The desire for climate control on a national scale has produced proposals ranging from the simply improbable to the terrifying.

Tantalized by the possibility of a Siberian breadbasket, the Soviet Union has broached some of the more spectacular schemes. One was to warm Siberia by damming the Bering Strait, thereby pooling the warmer waters of the Japanese Current off the coast of Vladivostok. This idea did not find favor with the countries on the other side of the strait. United States and Canadian scientists warned that diverted currents of Arctic water would flow across the top of North America, worsening the already uncertain climate of the Canadian maritimes. One suggested response on the part of the West was to install a series of nuclear heating plants in the Arctic Ocean above the proposed dam.

M. I. Budyko, a Soviet climatologist, suggested using nuclear-powered pumping stations to bring warmer water up into the Arctic Ocean and melt the polar ice cap. He calculated that once the ice was gone, polar temperatures would stabilize between 40 and 70 degrees Fahrenheit. Opponents in the United States claimed that a side-effect would be to submerge New York and London.

Several climate engineers have suggested plastic as a panacea against a new ice age. Plastic sheets suspended beneath the

ocean surface could modify evaporation rates and raise water temperatures. Plastic reflectors in space could warm icebound areas and increase the growing season in high latitudes.

While some scientists consider ways to manipulate climate, other analysts have concentrated on the political implications of natural cycles. In 1974 when a number of climatologists were warning of an imminent return to the ice age, the Central Intelligence Agency published two reports based on climatological predictions. They concluded that an ice age and the accompanying disruptions in tropical climates would give the United States "virtual life and death power over the fate of multitudes of the needy." The CIA rejected the possibility that the United States would abuse advantages inherent in this scenario, but it did suggest that countries with nuclear weapons might use

◄ Just over 10 percent of the Sahara is actually covered by sand dunes. Elsewhere the wind carries off lighter particles, leaving behind a stony surface called desert pavement, or *reg*.

them as blackmail on behalf of their hungry people.

It is safe to say that any actions directed at global climate, whether political or experimental, would have consequences unforeseen by their proponents. There are simply too many variables to fit into even the most sophisticated computer model.

Even small-scale modifications to an ecosystem are notoriously difficult to evaluate in terms of their effect on the larger climate. For example, villagers on the boundaries of the Sahara have pushed their land's carrying capacity to its limits in their struggle to make a living under marginal conditions. They overcut firewood and overtax supplies of ground water, while their livestock overgrazes the sparse foliage.

Increasing desertification is the result, but it is impossible to determine the exact point at which human actions become a cause rather than simply the hastening of an already inexorable cycle. Community aid projects such as tree planting and distributing solar cookers to conserve firewood may enable people on the fringes of the desert to restore a fragile ecological equilibrium. Ironically, the success of small-scale projects in moderating a desert climate has been inadvertently demonstrated in a community where survival is not the issue. Palm Springs, California, one of the nation's most affluent cities, has seen its average temperatures drop several degrees relative to neighboring towns. Researchers from Arizona State University in Tempe say the reason is golf. Since the early 1970s, Palm Springs has built more than 50 golf courses. Massive evaporation from the irrigated greens has served to lower temperatures throughout the city.

► The meeting place between land and sea takes many different forms. In Northern California, at 40 degrees north latitude, saturated air from over the Pacific Ocean forms mist as it rises over the hills along the coast. This moist temperate climate nourishes the huge sequoias of the coastal forest.

▼ On the west coast of Mauritania near Nouakchott, at 20 degrees north latitude, prevailing easterly winds carry approximately 250 million tons of sand and dust into the Atlantic Ocean each year.

and releasing heat-trapping pollutants into the atmosphere.

Even without human help, rain forests sometimes burn. The result illustrates both the vulnerability and the power of the tropical ecosystem. In 1983 a forest fire in Borneo became the biggest conflagration of the century. The fire began on the eastern side of the island, and then was fanned to immense size by the trade winds. It eventually consumed an area the size of Connecticut and covered the entire southern half of Borneo with smoke.

Just four years later, though, greenery had already made significant strides toward reclaiming the burnt ground. New trees were nearly 40 feet tall, and climbing vines ventured even higher up the skeletons of fire-killed giants. In the atmosphere, water vapor is once again rising from the leaves to

be returned as rain, and the fast-growing trees are reclaiming some of the carbon dioxide produced by the fire. Four hundred million years after plants first began colonizing the land, the urge of life to take advantage of empty ground remains as powerful as ever.

As we better understand the complexity of our atmosphere, we may also be on the way to realizing the organic unity of earth and air expressed in William Wordsworth's *Lines Composed a Few Miles from Tintern Abbey*:

"And the round ocean and the living air,
And the blue sky, and in the mind of man:
A motion and spirit, that impels
All thinking things, all objects of all thought,
And rolls through all things."

Here clearly is one of the central marvels of the planet.

◄ Riggs Glacier, Alaska. More than half the fresh water on Earth is in the form of ice. Photograph: Harald Sund.

►► Like flowing water, glacial ice gathers speed as its path narrows. This outlet glacier, squeezed between two mountains in the Juneau ice field in southeastern Alaska, can cover more than 15 feet a day.

RIDDLES OF SAND AND ICE

F OR most of us, ice is merely a seasonal
inconvenience. It may skim the surface of a
pond or complicate commuting to work on a
winter morning, but in the summer it is forgotten,
except for the little cubes that cool drinks.

Ice is much more than a regional curiosity,
however. Both poles of the Earth are locked in ice
all year, as are the highest peaks on every
continent. Even Africa's Mount Kenya, which is
located just south of the Equator, has several
glaciers above its forested lower slopes.

Ice is actually a complex crystal that varies
tremendously in appearance and behavior. Deep in
crevasses, glacier ice may glow like an eerie
sapphire, while chunks of sea ice — called *ivu*,
"leaping ice," by the coastal Eskimo — can
suddenly catapult themselves hundreds of feet
onto shore.

Sheets of ice have been advancing and retreating
across the globe for more than two billion years.
Their movements have created some of the Earth's
most striking topography and have affected regions
far beyond their own borders. Even today, the
winds that sweep the Sahara are part of a weather
system that is controlled to a significant extent by
the Earth's polar ice caps.

The last large-scale advance of glacial ice
occurred less than 20,000 years ago, just a moment
away in terms of the Earth's geological history.
Based on the cyclical changes in the Earth's orbit
around the Sun, we should now be entering
another long period of colder weather. Although
the Earth's atmosphere is presently warming as a
result of increasing carbon dioxide, the long-range
forecast is for cold.

A return to ice age conditions would spell the
end of human settlement in many of the world's

most populated areas. For example, what is now
New York state would once again be mostly
covered by glaciers. Ice once filled and polished
the floor of the trench that is now Long Island
Sound. It also scoured Manhattan Island down to
bedrock, making a firm foundation for the office
towers to come. In retreat, the glacier deposited the
oddly balanced boulders, called erratics, which
shade picnickers in Central Park.

The ice over Manhattan was at the edge of a
sheet that was as much as two miles thick over

▲ The towers of Manhattan could
be described as man-made
erratics, built of stone transported
from elsewhere, and stacked in
improbable formations on the
native rock.

▶ Transported stones like this boulder in Central Park were among the most persuasive evidence of an ice age to nineteenth-century geologists.

▼ Exposed by the force of a glacier, the schist bedrock of Manhattan Island supports a wilderness of buildings around Central Park.

parts of the upper Midwest and stretched almost as far as the site of present-day Indianapolis. Called the Laurentide ice sheet, it also bridged the Arctic Ocean between Canada and Greenland. Another ice sheet — the Cordilleran — spread out from mountains in British Columbia. Northern Europe carried its own blanket of ice, and the Antarctic ice cap extended north into the Southern Ocean.

Smaller ice fields were formed in Tasmania, New Zealand and South America. As the ice sheets grew, the oceans shrank accordingly, lowering the sea level by an estimated 350 feet. A land bridge united Alaska and Siberia, and what is now New York harbor was 100 miles inland.

The Laurentide ice sheet has advanced and retreated repeatedly over the past million years.

▲ The outer margin of the Laurentide ice sheet would all but bury the towers of Manhattan today. Further north, the ice completely engulfed the Adirondack and Catskills mountains.

► Summer meltwater trickles through a valley once filled by a river of glacier ice.

►► The track of a glacier is clear in this valley near Hudson Bay, scoured into a characteristic U-shape by the Laurentide ice sheet.

During this time the human species developed, evolving in the warmer latitudes and then dispersing all the way to the margins of the ice. The climate we know today is not the norm for our planet. Earth has been much warmer for most of its history. Neither is it a one-time aberration. The first documented glaciation took place about 2.3 billion years ago.

Glacial sediments from the first known ice age have been found in North America, South Africa, Australia, and India. These same deposits contain indicators of the arrival of free oxygen in the atmosphere, the result of widespread photosynthesis. Perhaps the burgeoning stromatolite colonies of the period absorbed enough carbon dioxide and emitted enough oxygen to dismantle the atmospheric "greenhouse" and help bring on the first glaciers.

The movement of continents through plate tectonics also affects the onset and intensity of ice ages. When there is a landmass in one or both polar regions, as there is today, ice fields form more readily and last longer. The action of the plates also affects the movement of the oceans which in turn affects heat circulation within the oceans, a major factor in global climate. Overlying these terrestrial events are perhaps the most important factors in the modulation of an ice age, the variations in the Earth's orbit around the Sun.

Few continents have entirely escaped glaciation. A landmass that was equatorial during one glacial epoch may have been subpolar during another. During the Ordovician period nearly 500 million years ago, what is now North Africa was near the South Pole. Huge boulders and glacial grooves, similar to the ones that point geologists toward Labrador as the source of the last great North American ice sheet, can also be seen in the stony plains of southern Algeria.

The ice sheets affect more than the ground they cover. Deep-sea core samples taken off the coast of Africa in 1986 show a correlation between high-latitude glaciers and equatorial marine life that can be traced back nearly three million years. Researchers in the Ocean Drilling Program found that the equatorial marine sediments were thickest — indicating abundant life — when the northern

► Neither Earth nor its orbit is a perfect sphere. The eccentricities of orbit and the tilts and wobbles of the planet itself are among the factors that influence the spread and retreat of ice.

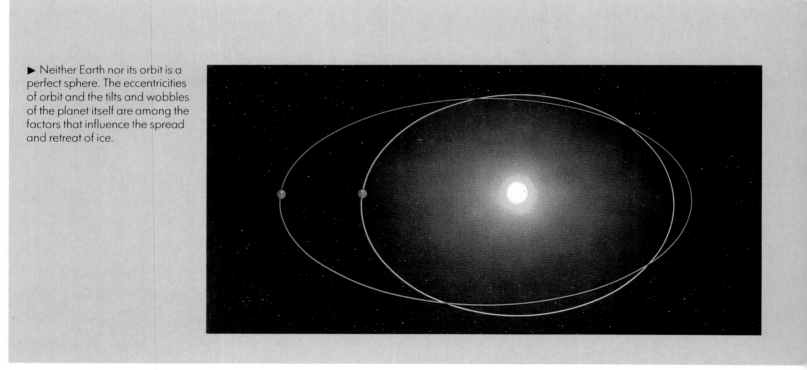

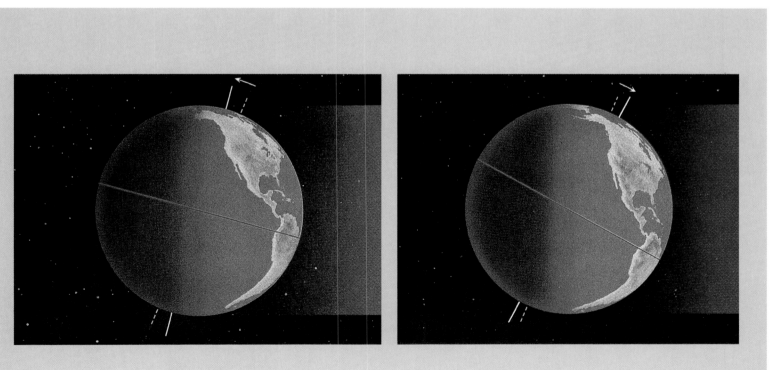

▶ Only the hardiest plants can survive in the dunes.

▶ All of Earth's climate is connected. This graph shows the correlation between warmer-than-average water temperatures in the northern hemisphere during this century and diminished rainfall in the Sahel region of Africa.

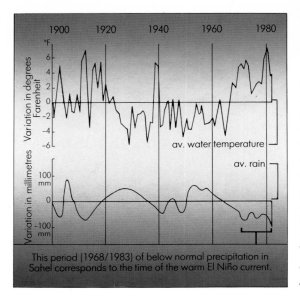

This period (1968/1983) of below normal precipitation in Sahel corresponds to the time of the warm El Niño current.

ice sheets were expanding. Samples from other locations show the same correlation. Either by the direct routes of winds and surface currents or by a more complex distribution system, cold water from the northern oceans found its way to the tropics, where upwellings rich in nutrients nourished a marine population explosion.

The interrelationship of ice, ocean, and air is also evident in the nature of the oxygen trapped in marine sediments. The oxygen in water, carbon dioxide and in the atmosphere has two isotopes: ^{16}O and ^{18}O. Although they are the same chemically, ^{16}O is lighter than ^{18}O. When marine organisms take up carbon dioxide and water from sea water to make their shells, they preserve a record of the ratio of ^{16}O to ^{18}O in the oceans of the world at that time. Marine animals that live in colder epochs deposit a higher proportion of ^{18}O in their shells and skeletons. These sedimentary archives aid in the difficult task of dating glaciations and determining the advance and retreat of glaciers during an ice age.

GLACIATION

To see an ice sheet at work in the present, scientists can go to Greenland. "A land that's seldom green," in the words of an old whaling ballad, central Greenland has been buried under ice for three million years. In an early example of real estate hyperbole, a tenth-century Norseman named Eric the Red gave the island its name in order to attract Viking settlers. For a few centuries their descendents in colonies along the southwest coast of the world's largest island found the designation accurate enough, but they and their settlement died out when the glaciers advanced again and their croplands froze. Only the Greenland Eskimo survived the cold and isolation of the Little Ice Age that descended in the fifteenth century.

Today Greenland's ice has receded once again, uncovering a fringe of coastal land. Settlers have reclaimed the new ground. They graze sheep in rocky pastures and steer their fishing boats amidst the icebergs of the Davis Strait. Not many venture inland onto the glacier where pristine blue rivers and lakes of meltwater dot the otherwise frozen landscape. Incredibly clear, these waters are created during the long hours of summer sunshine and return to ice when winter comes. They never reach the ground beneath the ice.

▲ At the terminus of a glacier, rocks ranging from boulders to fine-ground "flour" are left as the ice melts, in a pile called till, as seen here at Mount Cook, New Zealand. Photograph: L.E. Schick, Australian Picture Library.

ICE AGE

Is another ice age on its way? The intricacies of climate are such that it is nearly impossible to know for sure. Astronomical cycles point to a cooling trend (see page 190), but the atmospheric changes brought about by the unprecedented release of carbon dioxide since the Industrial Revolution may override the effects of orbit, at least over the next millennium.

In addition to these two indicators are a number of less predictable agents of long-term climatic change. The amount of energy from the Sun might fluctuate, or something might keep the radiation from being absorbed. One possible sunscreen is an increase in terrestrial cloud cover, although the same clouds that block incoming radiation can also block outgoing radiation, thus muting or even reversing the cooling effect.

Volcanic debris and gases can reduce sunlight for months and even years. Depending on the season, it may either lower or raise global temperatures. The sulfuric acid that is a variable component of erupted material may be a more potent cooling force than dust. Droplets of sulfuric acid have been found along with other volcanic debris preserved in glaciers, and correspond well to periods of glacier growth.

It would take an extraordinary eruption to cause the eight to ten-degree drop in Fahrenheit readings that would probably trigger an ice age, but a volcano or a meteorite collision might combine with other factors to turn snowfields into glaciers. The eruption of Mount Tambora in 1816, which led to the "year without a summer" in the northeastern United States and Northern Europe, dropped global temperatures an estimated 3.5 degrees Fahrenheit. Summer frosts ruined the potato harvest in Ireland, and the resulting famine led to a typhus epidemic that killed 65,000 people. This period of unusual cold was followed, however, by a marked warming trend that persisted into the mid-twentieth century. Subsequent eruptions have temporarily interrupted the rise in temperatures, but none has had enough impact to reverse it.

Changes in the Sun itself are harder to evaluate, but there do seem to be peaks in solar output. Some physicists suggest that the Sun waxes and wanes in size due to the pattern of burning in its core. When core burning is expanding, more of the Sun's energy is used on its own processes and thus never reaches Earth. This cycle, lasting millions of years, may trigger the great ice ages. Shorter glacial fluctuations could be attributed to "flickering" caused by irregular mixing of matter in the Sun.

In the late 1970s the specter of an imminent ice age occupied the same place in newspapers and popular magazines that was held in the late 1980s by warnings of global warming through the greenhouse effect. One of the most urgent alarms was sounded by popular British writer Nigel Calder in 1978. Calder envisioned a chain reaction where one cold summer begets an ice age within a decade.

Glaciologists did not concur with Calder's scenario, which had 15 countries in danger of vanishing under the ice, and another 20 suffering from encroaching glaciers. And in fact, since Calder's season of celebrity, the world has shown signs of heating up. The four warmest years of the past century occurred since 1979. While scientists continued to forecast an ice age 20,000 years hence, popular attention shifted to the dangers of

global warming. The carbon dioxide build-up underway for decades suddenly drew the attention of the United States Congress and other governments, and the greenhouse effect became a household term.

However, similar constellations of warm years have occurred in the past without noticeably affecting long-term trends, and some climatologists are convinced that a real manifestation of the greenhouse effect would show a different pattern. According to the most accepted climate models, global warming should show up first in polar regions, while the heat waves of 1987 and

1988 have been concentrated in mid-latitude zones like the American Midwest.

Furthermore, the Earth's own feedback systems may alter the effect of increased carbon dioxide in unpredictable ways. Increased evaporation due to warmer temperatures could mean more precipitation and less sunlight penetrating the cloud cover. These cooling mechanisms might be enough to offset or even overcome the increased insulating ability of the atmosphere.

If so, ice that is now in retreat may begin to push its way back down the mountains and into the headlines.

▲ Niagara Falls were carved out by a glacier. Since then erosion has been pushing the cataract north toward Canada.

Radar measurements from the Greenland Geological Survey' show that one-third of the island's rocky base actually lies below sea level. The weight of ice has depressed the rock more than 2,000 feet in some places, displacing the Earth's crust into the more malleable mantle beneath. When the ice sheets retreat, as they have in the current interglacial era, the rock rebounds. Its rise can be tracked through the series of dry abandoned beaches left behind, and its slow progress can be measured by using laser beams reflected from space. The area around Hudson Bay has experienced the greatest lift in North America. The Hudson Bay area has already emerged nearly 1,000 feet, or about half its preglacial altitude, and is still rising.

On small plots of ground, the effects of this rebound can be especially dramatic. Replot Skelly Guard, a group of small islands off the coast of Finland, have increased their land area by more than one-third since 1800. Islanders are raising vegetables where their grandparents fished for cod. The rising landmasses would be even more noticeable if sea levels did not also rise, fed by the melting ice.

AN ICE SHEET AT WORK

Toward the end of the Little Ice Age in the mid-nineteenth century, a global warming trend sent Alpine glaciers melting back up their valleys. The Rhône Glacier in Switzerland had retreated more than a mile and a half by 1920. It was followed by amateur and professional glaciologists every step of the way. But when the climate began to cool again in the 1940s, glaciers responded promptly. Even the warming trend of recent years has not sent all glaciers into retreat. The Hubbard Glacier near Yakutat, Alaska, is among the ice fields that is now increasing its range, although not because of global climate changes. Hundreds of marine animals were cut off from the sea when it closed the entrance to a fjord in 1986, and the rescue effort that followed drew international attention.

◄ Chunks of ice jostle fishing boats at Jakobshavn Sound in Greenland. The supply is replenished by Jakobshavn Glacier, which discharges 20 to 30 million tons of ice a day to the fjord at its mouth.

▲ Water from a seasonal lake on the Greenland ice sheet tumbles into a crevasse.

▶ Two tributary glaciers join to form a river of Greenland ice. Rocks and debris at the margin of the flows create the dark stripe known as a medial moraine.

Citizens of Juneau, Alaska, took a particular interest in the events at Yakutat. Mendenhall Glacier has not expanded in recent years, but it still reaches within five miles of the state capital's airport. Boaters on Mendenhall Lake at the glacier's foot are warned against hypothermia and sudden noises — the first for obvious reasons, the second because sound waves can dislodge tons of unstable ice from the blue-veined cliffs that rise 100 feet above the water.

Twelve miles back from the lake is the Juneau ice field, source of 36 glaciers including Mendenhall. An expanse larger than the state of Rhode Island, the ice field is nearly a mile thick in

places. Solitary mountain peaks termed *nunataks* — an Eskimo word meaning "lonely peaks" — rise like islands above a sea of white. Though much smaller, the Juneau ice field is formed by the same processes that created the gigantic Cordilleran and Laurentide ice sheets. Beginning one summer when the winter's snow pack does not melt away, snow begins to accumulate. The bottom layers are compressed by the weight above, which can create pressures of up to three tons per square foot at a depth of 100 feet. The scattered snow crystals gradually recrystalize into compact grains of ice. If the pressure continues to build, the grains form a body of ice which starts to slide. The snow field has become a glacier.

▶ The terraced slope above Hudson Bay is still rebounding from the weight of the Laurentide ice sheet, gaining about two feet of elevation per century. Each rocky ledge marks a previous shoreline.

▼ Radar penetrates ice with ease, an attribute that has misled many pilots into catastrophe. It also means, however, that the long-covered rock of central Greenland can be mapped. More than a third of the island's surface has been pushed below sea level by the weight of ice.

EVOLUTION OF A GLACIER

The evolution of a glacier can be seen in the deep layers exposed by crevasses. Beneath the snow of the previous winter is a transitional layer of *firn,* from the German word referring to "last year." In *firn,* the air pockets that give a snowfall its magical silencing effect have been filled by meltwater that trickles down from the surface during warm spells and refreezes in the interstices. The formerly lacy hexagonal crystals have been rounded and compressed by pressure.

As the weight of accumulating snow and ice increases, the *firn* is compressed into denser and denser forms of ice, designated by glaciologists as ice one through ice eight. Although all are ice, they differ measurably in certain basic physical attributes, such as specific gravity. Dark horizontal

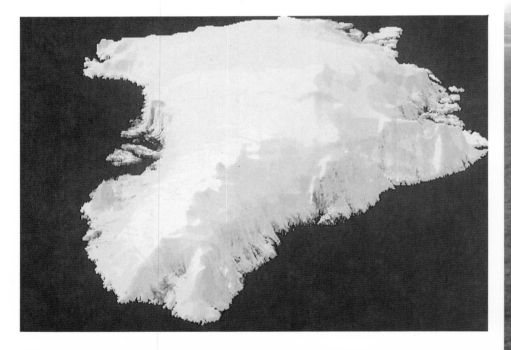

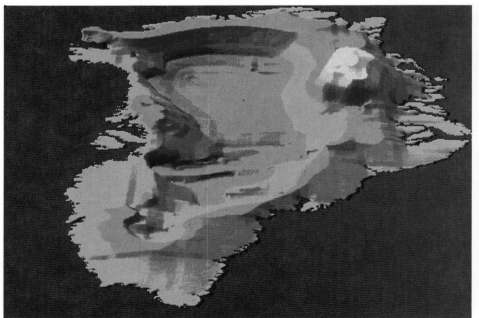

stripes in the *firn* mark annual accumulations of snow. The dark layers come from the dust and debris that accumulate on the glacier's surface in the summer. This litter can be preserved for thousands of years, only to be revealed when glacial ice splits into a crevasse. Its analysis yields valuable information about the environment of its time, including traces from fires, volcanic eruptions, and even plant pollens.

Crevasses may be more than 100 feet deep. Beyond that depth the ice is no longer brittle enough to split, so glaciologists must make their own entrée. To reach the point where ice meets bedrock and the real work of glacial excavation takes place, researchers from the University of Washington cut a 300-foot tunnel through the side of Blue Glacier on the Olympic Peninsula's Mount

▲ Students from Nepal, China, and West Germany, as well as Canada and the United States, come to the Juneau ice field to study glaciers. The research camp is located on the more stable snow at the base of a mountain.

◄ Dr. Maynard Miller of the University of Idaho is among the researchers studying the transition from fresh snow to glacier ice.

◄◄ Set on a narrow strip of land between coastal mountains and the North Pacific, Juneau, Alaska, has spread to within a few miles of Mendenhall Glacier.

LOUIS AGASSIZ AND J H. BRETZ

The legacies of the great ice sheets — including moraines, erratics, grooves — are obvious once the eye is informed. Even beyond the range of the tens of thousands of glaciers still flowing throughout the temperate world, it is easy to forget that an ice age seemed as wild an idea as continental drift or the big bang to those who first heard it.

The French naturalist Georges de Buffon suggested in 1778 that the Earth was cooling and that life forms had retreated from their former homes at the poles. Early in the next century a Swiss mountaineer and a Connecticut manufacturer both began telling friends that the odd boulders near their hometowns had been deposited by glaciers.

But the ascension of ice age theory is credited to Jean Louis Rodolphe Agassiz, an energetic Swiss scientist who saw the light during a summer vacation in the canton of Vaud in 1836. The moraines of the Rhône Valley and the glaciers of Chamonix made him a passionate advocate of L' Epoque Glaciaire. "Since I saw the glaciers I am quite of a snowy humor," he wrote to a friend, "and will have the whole surface of the earth covered with ice, and the whole prior creation dead by cold."

Agassiz did not mean this great annihilation as hyperbole. A meticulous observer and a world expert on fossil fish, he was also a passionate believer in what is now called creation science. There was no evolution in his world view. Species were wiped out by divinely ordered catastrophes, and then God created new ones. An ice age was simply "God's great plough," a more scientifically plausible version of the biblical flood. This doctrine of special creations was widely believed — it would be another 20 years before Charles Darwin published the *Origin of Species* but the most respected figures in geology had begun to reject dramatic cataclysms in favor of slow, steady, uniform processes.

"We are not authorized in the infancy of our science to refer to extraordinary agents," wrote Charles Lyell in 1833. Elsewhere in *Principles of Geology,* the discipline's seminal text, he complained that "We are . . . told of general catastrophes and a succession of deluges, of the alternation of periods of repose and disorder, of the refrigeration of the globe, of the sudden annihilation of whole races of animals and plants, and other hypotheses, in which we see the ancient spirit of speculation revived, and desire manifested to cut, rather than patiently to untie, the Gordian knot."

Still, Lyell became an early convert to the idea of an ice age. The fact that glaciers could still be seen at work suited them to his belief in uniformitarianism, the belief that the forces of the past are still at work in the present. Darwin also wrote to praise Agassiz for his research. When Agassiz moved to the United States in 1846, his reputation as an original thinker preceded him. From his first steps off the boat at Halifax, he delighted in finding "the polished surfaces, the furrows and scratches, the *line engravings* of the glacier." He settled in Boston, where he became a popular Harvard professor, a museum director, and a national icon of science.

For the rest of his life, Agassiz searched the New World for signs of ice and found them everywhere. Most of his discoveries were authentic; some, like his announcement of recent glaciation in Brazil, had more to do with his preconceptions than with the evidence. Regional glaciations were not enough for him. In order to preserve his belief in a separate creation, he had to find confirmation that the killing frost was worldwide in scope. During his life, his opinions were advanced by his own celebrity. When he died in 1873 they encountered more criticism. His successor in the chair of geology at Harvard went so far as to encourage his colleagues to reject all ice age theories "without hesitation."

The evidence was piling up, however. Shortly after Agassiz's death a find in the Midwest established a record of more than one ice age. The remains of a forest were found between two layers of glacial sediment. The quest for glacial evidence moved into a new dimension as geologists tried to establish not only how far the ice had advanced but how many times it had been there. Eventually they were able to push the earliest of the ice age cycles back more than two billion years. Even Brazil was found to have traces of former glaciation, though not from the era Agassiz had claimed.

The triumph of the ice age as a geologic explanation did not mean general acceptance of catastrophes as agents of change. Science — especially Earth science — has remained largely wedded to uniformitarianism. As proponents of meteorite extinctions have found, anyone who proposes a cataclysm faces a battle based on emotion as well as analysis. A case in point is the debate over the channeled scablands of the Columbia Plateau. Across the Cascade Mountains from the green valleys and crowded freeways of western Washington, the Columbia River winds through a landscape sculpted by fire and ice.

The basalt formations that characterize the Columbia Basin were laid down in a series of fissure eruptions beginning about 15 million years ago. Layers of lava averaging 100 feet thick built up the land inside a circle of mountains. Between eruptions the basalt was covered with river sediment. Many trees, including gingkos, grew in a semitropical climate, sheltering giant rhinos and camels. By the time the weather changed and the Cordilleran ice sheet spread south, the northern ice was effectively funneled across the lava plateau.

J H. Bretz, a geology professor at the University of Chicago, proposed in the 1920s that the coulees and channels were not created bit by bit, but suddenly in the course of one tremendous flood. He could find no other explanation for the erosion patterns and sediments left behind. To most of his peers, the "Bretz floods" seemed biblical in implication as well as in size. Dedicated to uniformitarianism, they derided his observations and held out for slow and steady erosion by rivers.

It turned out to be a conflict of principle versus observation, and Bretz was the better observer. Eventually, a source was found for the water he hypothesized. Glacial Lake Missoula was formed when a lobe of Cordilleran ice sheet blocked the Clark Fork River in the Idaho Panhandle. Periodically the ice dam failed, and as much as 50 cubic miles of water rushed west across the plateau. The huge lake probably drained in less than two weeks, overflowing the river channels and creating the vast scablands of the Columbia Plateau.

Aerial photographs and satellite imagery have further confirmed Bretz's earthbound observations. They show giant ripples on channel floors that could only have been made by

massive quantities of flood water in a hurry. Detailed studies of the scabland sediments indicate not one flood but several. Nevertheless, Bretz's insight has stood the test of later analyses. In 1976, when he was 95 years old, Bretz was given geology's highest professional honor, the Penrose Medal of the Geological Society of America. The facts of his catastrophe triumphed over the theory that denied them.

▲ Channeled scablands of the Columbia Plateau are the result of the floods that washed away river sediments and exposed volcanic basalt. Photograph: Harald Sund.

► Researchers collect samples from a test hole drilled high in the Juneau ice field. Shirtsleeves and sunscreens replace parkas on the rare clear days of the coastal summers.

► The Juneau ice field in southeastern Alaska was once part of the giant Cordilleran ice sheet that covered what is now western Canada.

Olympus. The films they made show the process by which some coastal glaciers can turn a shallow valley into a deep fjord.

A glacier's power comes from a combination of weight and abrasion. Rocks ranging from boulders to fine-ground "flour" travel with the ice. They are the teeth of the glacier, gouging the bedrock as they pass.

At the glacier's terminus the rocks are left in an unsorted pile called till as the ice melts. Linear ridges of debris at the glacier margins are called moraines. The moraine bounding Mendenhall Lake

marks the extent of the previous glacial advance, and serves as a dam for the meltwater that fills the lake.

MELTWATER

Meltwater also carries a load of debris, but unlike ice, it sorts its load by weight. Larger stones stay close to where the ice left them, while finer rocks and sand travel along the watercourse. The sandy and gravelly deposits that result are called "outwash plains." On Long Island, for example, the

► This erratic was carried westward from the Rocky Mountains to an Alberta wheatfield. Ice sheets also quarried gold and even diamonds from the ancient rock of the Canadian Shield, and deposited them as far south as central Indiana.

►► A natural crevasse on the Juneau ice field. The uppermost layer—20 feet deep—is the previous winter's snowfall. The dark stripes beneath are annual accumulations of dust and other debris, marking off the transition from snow to *firn* to ice.

Flatbush area of Brooklyn is a terminal moraine marking the glacier's farthest reach, while the beach at Coney Island is the huge outwash plain. The lightest particles of glacial flour are carried the farthest, turning meltwater rivers a characteristic milky white.

Once the signs of their passage are known, the tracks of glaciers are evident throughout the higher latitudes. The Matterhorn and its neighboring Alpine spires were whittled by constellations of cirque-cutting glaciers, which plucked away at the rock as they passed. The Great Lakes were ordinary river valleys until ice from the Laurentide ice sheet gouged them down to 700 feet below sea level. The stone fences and rocky soil of New England are products of glacial deposition. The boulders that rise from the wheat fields of Alberta and along paths in Central Park are non-native rocks that were carried along by the ice and left behind when it retreated.

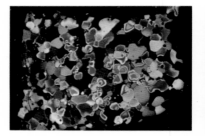

▲ Seen under a polarizing plate, the crystal shows a rainbow of facets. The black areas within the crystal are air pockets, indicating that this crystal is from relatively near the surface.

► Summer at Mendenhall Lake. The pinkish areas are colonies of a species of algae that thrives on glacier ice.

LANDSCAPE OBSTACLE COURSES

Yet another relic of the glacier, as Dr. Anita Harris points out in John McPhee's exploration of American geology, *In Suspect Terrain,* is the golf course. Golf was invented in Scotland, where the landscape emerged from the ice with the hummocky moraines and outwash sand patches that combine to make the perfect natural obstacle course. In places where ice has not left its mark,

developers employ bulldozers to act as instant glaciers.

Following the clues left in the rock, we can follow the Laurentide ice sheet back to its source. The bedrock in Central Park is marked in parallel lines by the rocks at the bottom of the most recent glacier. A compass reading shows that the glacier ice that carved them travelled from the northwest toward the southeast. In upstate New York, the Finger Lakes are also broadly aligned along the

▲ Recent deposits of terminal moraine at Mendenhall Lake.

◄ The repeated freezing and melting along a glacier's surface plucks away at the mountains in its path, eventually tearing off enormous chunks of stone, like this one, and leaving behind the characteristic cirques and spires of glacier-melted ranges.

187

glacier's path. Here the soft shale bedrock has been burrowed, as if by the fingers of a giant hand. The largest of these 11 lakes, Cayuga, is nearly 40 miles long.

Further north, the great north woods thin into a scattering of stunted conifers before giving way to tundra. Small lakes dot the land from horizon to horizon. Although they differ in shape from the

Finger Lakes, they also were created by the ice sheet. Seen from the air, they give the land a tiled pattern reminiscent of the overlapping of a bird's feathers. Deep grooves of the sort seen in Central Park are common here, as are erratics. The largest of these stones weigh hundreds of tons, and a surprising number are precariously balanced. This is a distinctive sign of ice, for unlike floods,

◀ "Snow cats" transport glaciologists to and from their camps on the Juneau ice field. Rough terrain and hidden crevasses make glacier travel especially perilous.

glaciers often deposit their debris gently, almost like falling snow.

LABRADOR-UNGAVA PENINSULA

At the heart of this bleak landscape lies a subarctic tableland called the Labrador-Ungava Peninsula. Today, the winter snows that cover the peninsula melt away briefly in the long days of northern summer. For most of the last 100,000 years, however, this country has been the breeding ground of glaciers that helped create nearby Hudson Bay and left the peninsula largely devoid of topsoil. The glacial grooves extend into the bedrock of the Canadian Shield, the most ancient portion of the North American continent.

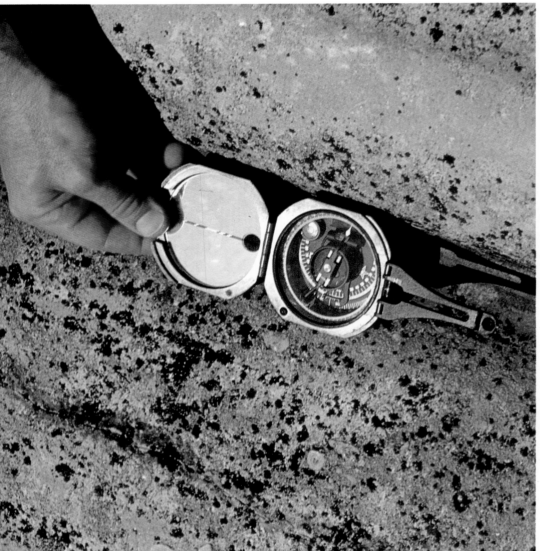

▲ Kelley's Island in Lake Erie shows the same alignment of glacial grooves as Central Park. Long before they knew their origin, dwellers along the route of ice used grooves like these as a compass.

▶ Dr. Robert Ridky, at left, retraces a glacier's path across Kelley's Island. The island's limestone surface is marked by grooves hundreds of feet long and up to 15 feet deep.

At its most basic, an ice age is a simple phenomenon. All that is needed is cold weather and water. In practice, the forces that drive such a drastic alteration in the Earth's weather are a complex mix of terrestrial and extraterrestrial forces. On Earth, rising concentrations of carbon dioxide and other "greenhouse" gases lead to warmer weather. Conversely, meteorite impacts and volcanic eruptions can throw enough debris into the stratosphere to dim the effect of the Sun and lower global temperatures for months. Events on the Sun also can affect the amount of radiation the Earth receives, making the climate warmer or colder.

THE EARTH'S ORBIT AND CLIMATE

However, the most crucial cycles governing the advance and retreat of glaciers appear to be functions of Earth's movement through the solar system. The theory correlating ice ages with fluctuations in the Earth's orbit is named for Milutin Milankovitch, the Yugoslav physicist whose calculations helped refine it in 1941. Milankovitch

▶ Some erratics on the Labrador-Ungava Peninsula have been carried more than 800 miles. Eight thousand years after the ice made its last major advance, the exposed Canadian Shield supports little more than a somber-hued collection of lichens.

took as a starting point the conclusions of James Croll, a self-taught Scottish scientist who started his research on ice ages in 1864 while working as a janitor in Glasgow.

Croll and Milankovitch concentrated on cyclical variations in the Earth's orbit as an agent of climatic change. From the time of Kepler, astronomers had known that the Earth's orbit around the Sun is elliptical, not round. Furthermore, it is constantly changing, becoming more and less elliptical in a 100,000-year cycle. The distance between Earth and its star can vary by more than 11 million miles, with a corresponding change in the intensity of solar radiation in any given hemisphere and season.

The tilt of the Earth on its axis also varies according to a cycle of about 41,000 years. It is the tilt that creates the seasons, as first one hemisphere and then the other has long days and short nights. If the Earth stood upright on its axis, every day would be as long as every night. As the tilt moves from 22 degrees to its maximum of over 24 degrees, seasonal differences become more extreme. At 24 degrees, a northern region like the Labrador-Ungava Peninsula receives more sunshine than usual and so may experience warmer weather. When the angle is smaller, the climate tends to be cooler. The tilt today is 23.4 degrees and getting smaller.

Because it is oval rather than completely round, the Earth responds to the gravitational pull of other bodies and to the differing pressures of its own atmosphere by wobbling on its axis like a top. There are several cycles of wobble, ranging from about 23,000 years to as little as two weeks. The most influential in terms of climate are the 19,000 and 23,000-year-cycles, the latter of which has the stately title of "precession of the equinoxes."

Eleven thousand years ago, when the Laurentide ice sheet was rapidly retreating, the planet was tipped so that the northern hemisphere was closest to the Sun in summer and farthest away in winter. The aphelion — the day when the planet is at its

greatest distance from the Sun — came in January. Today the situation is reversed, which contributes to cool summers in the northern hemisphere.

These overlapping cycles of orbit, tilt, and wobble cause considerable variations in the distribution of solar heat. By plotting their combined effects, Milankovitch was able to establish how much solar radiation a particular latitude had received at any given period. Later scientists matched these levels with the dates of known ice ages. The correspondences they found are not perfect, but they are close enough to give the theory credence.

Cool summers rather than fierce winters seem to be the crucial element in the start of an ice age. Milankovitch concluded that the most critical variable was the amount of radiation received at 65 degrees north latitude, slightly north of the Labrador-Ungava Peninsula. Snow that does not melt there during the summer prepares the way for more snow. Its white surface reflects solar radiation, reducing evaporation and insulating the ground it covers. When the next winter comes, new snow begins to compress the layers beneath into *firn*. A series of cold summers can start a sort of chain reaction, as chilled air above the snowfield squeezes the moisture out of the passing clouds and brings yet more snow.

When glaciers retreat, the same kind of reinforcing cycles work in reverse. A series of warm summers reduces the glacier's margins. Uncovered, the darker ground soaks up radiation and transfers heat to the air above, hastening the melting of more ice. The whole world feels the change as temperatures rise and water that had been locked up in ice returns to the ocean. Even latitudes that never see a snowfall are still affected by alterations in patterns of wind and rain.

CLIMATE AND THE SAHARA

Dramatic evidence of this can be seen deep in the Sahara desert at a place called Tassili N'Ajjer. The name Tassili N'Ajjer means "plateau of flowing

▼ No climate on Earth is constant. Beneath the dunes of the central Sahara lie ancient river courses and flood plains, some dating back more than two million years.

◀ Drought has decimated many settlements in Mauritania, forcing families to abandon homes they had laboriously wrested from the desert.

▼ A cooperative farm in Mauritania. The same irrigation that produces vegetable crops contributes to a loss of groundwater and eventual desertification.

◀ Established in 1968, and replenished by the rains of the mid-1970s, this well can no longer support the people and livestock who depend upon it.

◄ The sandstone spires of Tassili N'Ajjer in the central Sahara are the eroded legacy of the sea that covered the area 130 million years ago.

▶▶ A pack train carrying supplies crosses the crumbling stone at the base of Tassili N'Ajjer. Even in this arid land, it is primarily the action of water that gradually transforms mountains into rubble and then to sand.

rivers," and indeed many old river courses and canyons can be found in the rock. Even more dramatic evidence of water in the Sahara lives in the canyon's shade. Here a handful of cypress trees that may have sprouted over 4,000 years ago still spread green among the blazing red rock. They date to a period when the Sahara received bountiful rains.

Although the people themselves have been gone for millennia, the residents of the ancient Sahara left records of what the area was like in the days when trees and beasts flourished at Tassili N'Ajjer. Between 6,000 and 2,000 years ago, they carved and painted thousands of images from their lives on rock walls.

The most recent pictures show camels, which suggests a dry climate like the one today, but earlier wall painters depicted different animals. Perhaps wishing for an abundance of livestock, ancient farmers painted many cows around them. Elsewhere we see cattle grazing, people cooking and chopping wood. The oldest paintings show nomadic hunters pursuing beasts which no longer inhabit this area.

What caused this dramatic change in the North African climate? The climatic changes witnessed at Tassili N'Ajjer are driven by fluctuations in the two great "dry belts" that circle the globe above and below the Equator. These dry belts — known to sailors as the horse latitudes — are caused by the same global air circulation systems that are responsible for the trade winds and jet streams. The location of the cells is a function of the temperature changes between tropical, temperate, and polar regions. When the polar ice sheets expand these boundaries also shift, relocating the dry belts.

Around 18,000 years ago, glacier ice reached its maximum advance, especially in the northern hemisphere. The dry belts in the northern hemisphere were pushed south toward the Equator. This gave areas in the northern Sahara like Tassili N'Ajjer a wetter and more moderate climate, similar to what the Mediterranean enjoys today. At

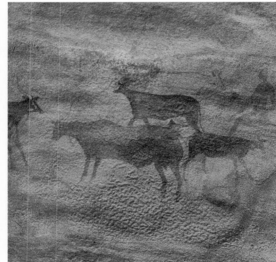

▲ Lifesize images incised into a sandstone cliff at Tassili N'Ajjer recall a time when inhabitants had enough water to raise cattle and enough leisure to celebrate their good fortune.

◄ The middle range of pictures feature herdsmen and their cattle. Cattle need more forage and more water than camels.

◄◄ The most recent of the more than 4,000 paintings found at Tassili N'Ajjer feature camels, an indication of increasing aridity.

◄◄ A Tuareg tribesman, dressed for protection against the heat and dust, points out one in a series of superimposed rock paintings.

◄ Antelope like these roamed the now-vanished grasslands of the Sahara.

◄ As their life became more settled, the herdsmen of Tassili N'Ajjer began to commemorate home and hearth.

▲ The lush shoreline of Lake Chad is deceptive. The lake is shrinking so fast that these plants could be desiccated in a matter of months.

▶ Right and far right: Chad is one of the world's poorest countries. Many of its people depend almost totally on the fish from Lake Chad.

the close of the last ice age the southern Sahara still was hospitable, but the drying trend was underway. Drought began to wither the vegetation and use up the ground water.

In historical times, the southern limits of the desert have fluctuated — first north, then south — occasionally bringing severe drought to inhabited areas. Today, the Sahara is expanding southward at an accelerating clip. In Mauritania, dunes have overwhelmed thousands of farmers and herdsmen and driven them into massive refugee camps around Nouakchott.

Meanwhile, the desert has nearly dried up the southern Sahara's largest lake. In 1973, Lake Chad was as big as the state of New Jersey. Rich in fish, it was a major source of sustenance for one of the poorest regions of Africa. Today, Lake Chad has shrunk to less than a tenth of its size a decade ago.

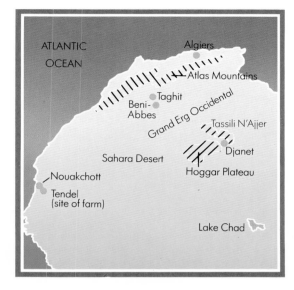

► Six weeks before this photo was taken, this area was under water. A narrow channel is all that remains.

►Villagers join forces to haul boats to the lake, a job that used to take minutes and now can take as much as four days.

Lake Chad and the desert around it are being shaped by global weather patterns, some of which can be traced to the planet's polar ice. Today, two contradictory influences are at work on the Earth's climate. Astronomical cycles are making the Earth colder; human activities are making it hotter.

Analysis of natural changes in the Earth's orbit shows that we can look forward to 60,000 years of adverse orbital geometry. There will be less and

less warmth from the Sun reaching the Earth just as surely as if the thermostat had been turned down.

Yet the Earth is showing signs of a short-term warming trend attributed to carbon dioxide and other greenhouse gases building up in the atmosphere. For years — no one knows how many — the greenhouse effect is likely to override the changes in orbit.

Only time will tell what climate this combination of forces will create.

▲ Villagers must haul their boats farther and farther to reach the receding shoreline. Some families have been able to grow vegetables in the moist topsoil exposed by the vanishing lake.

◄ Near Pokhara in the Himalaya Range of Nepal. Even the most remote places on Earth are influenced by human actions.

►► Center pivot irrigation in a Nebraska cornfield. The amazing yields of the mechanized American farm have come at a high cost in global resources.

THE HOME PLANET

FROM space, the surface of the Earth shows the changing patterns of a living planet. Clouds and smoke weave delicate textures over land and sea. The deep green of tropical forests contrasts with the tan of savannas and the white of polar ice.

Signs of human civilization are also evident. In the Sahara desert, a strange pattern of interlocking circles interrupts the austere expanse of sand and rock. Each circle is a field that is irrigated with water drawn from underground wells. On the far side of the globe, similar mile-wide circles mark individual irrigated plots in the desert of eastern Oregon. In many areas of South America's Amazon Basin, the most obvious surface feature is deforestation. Satellite images show that large areas of the Amazon have been devastated along a branching network of logging roads that push ever deeper into the vanishing woods. The rocketing world population — which reached 5 billion in January 1987 — means human influence on the planet will continue to increase.

HUMANS AFFECT THE ENVIRONMENT

Human beings are by no means the first, or the only, species to affect their environment. Every living thing contributes to the biosphere in some way. Some, like the pioneering photosynthesizers, transformed the planet profoundly and caused the extinction of many other organisms in the process. Humans have come late to the game, but we are unique for the speed at which we have developed into a global force, as much as for our capacity for reflective intelligence. No other species aspires to be the crown of creation, or tries to leave a record of its progress.

Early evidence of humankind's keenness of mind and eye survives from the height of the last ice age. Nearly 30,000 years ago, hunters painted now-extinct beasts on the walls of southern European caves at Altamira and Lascaux. Even though their stone tools were crude, the people who painted these images possessed the intellectual and technical expertise to create some of the most powerful art the world has ever known. In addition to painting mammoth, ibex, bison, bear, horse, and reindeer, the Cro Magnon artists

▲ Without irrigation, much of eastern Oregon is too dry for crops. With water systems like these, drawn from the Columbia River and its tributaries, it produces wheat, potatoes, and other crops.

▲ The limestone caves of the Dordogne region of France were made by the same process as the caves of Guilin, China. Beginning at least 20,000 years ago, they served as shelters and galleries for the human societies that colonized Europe on the fringes of the last ice age.

▶ This moose is from a cave near Cognac, France.

▶▶ The famous walls of Jericho now enclose an archeological treasure, a community dating back nearly 10,000 years. The remains of the palace of Herod the Great also have been excavated a few miles away.

sometimes left their signatures in the form of hand prints.

At that time, the total world population was probably no more than 5 million. Clothed in furs and armed with stone-tipped spears, the people lived a meager existence in the shadow of the massive ice sheet that covered much of western Eurasia. Even then, though, they were launched on the road that would shape the face of the Earth to a

startling extent in just a few thousand years. During the Neolithic period that coincided with the warmest part of the present interglacial epoch, tribes began to give up nomadic hunting and herding to establish their first settled communities. One of these is ancient Jericho, near the Dead Sea in what is now the West Bank area of Israel.

Recent archeological excavations have revealed part of the storied walls of Jericho. The sack of

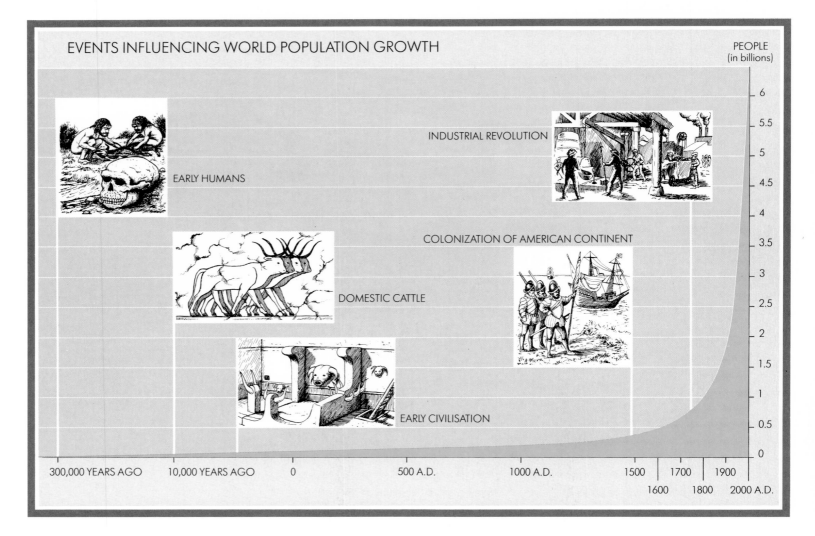

EVENTS INFLUENCING WORLD POPULATION GROWTH

PEOPLE (in billions)

EARLY HUMANS

INDUSTRIAL REVOLUTION

COLONIZATION OF AMERICAN CONTINENT

DOMESTIC CATTLE

EARLY CIVILISATION

300,000 YEARS AGO   10,000 YEARS AGO   0   500 A.D.   1000 A.D.   1500   1700   1900

1600   1800   2000 A.D.

6
5.5
5
4.5
4
3.5
3
2.5
2
1.5
1
0.5
0

▲ Female figurines like this one from Jericho also have been found in many older sites in Europe. They are thought to be fertility symbols.

◀ A mask from Jericho.

◀◀The ruins of a school in an abandoned Mauritanian village. More than half the population of the country's interior has been forced to leave.

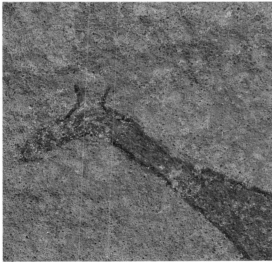

▲ Jericho was an agricultural city, and this mortar and pestle was used for grinding grain.

◀ A giraffe from the caves of Tassili N'Ajjer, from the same era as the sculptures of Jericho.

Jericho depicted in the Old Testament was only one of many times the city was destroyed and rebuilt. Scholars estimate that Joshua's conquest of Palestine occurred about 3,400 years ago, but the oldest known level at Jericho is dated closer to 10,000 years old. The Hebrew translation of Jericho — City of the Moon God — is another indication of the city's antiquity.

Judging by the finely crafted stone sickles found at Jericho, it appears that the town's economy was based on grain production, probably in the fields outside the town walls. The presence of mortars and pestles for grinding grain underscores the economic link between agriculture and a town economy: even if they come upon a field ripe for harvest, a nomadic people cannot transport the technology to grind and store grain. It is estimated that 2,000 people lived in ancient Jericho, making it a metropolis of the time.

Similar agricultural communities spread all around the Mediterranean, including areas that now are part of the Sahara desert. Even in the central Sahara, clear evidence remains that people led an agricultural existence thousands of years

ago. On the rock cliffs, they skillfully carved images of their cattle. Wild animals are depicted as well, including a giraffe rendered life size.

In those days what is now the world's largest desert was probably a savanna with intermittent stands of trees. Desertification was set in motion

213

by the shifting of dry belts at the end of the last ice age, but humans also have played a role in loosing the sands of the Sahara. As the rains fell less and less often along the North African coast, the desert spread south, west, and east.

As forage gradually withered and died, more and more formerly nomadic cattle herders were forced into river valleys such as the Nile. The first Book of Moses contains an account of the successive migrations of the ancestors of the Israelites in a dying landscape that probably refers to this process.

Along the Nile, this influx of immigrants helped provide both the skilled craftsmen and the slave labor that built the world's first great civilization, the Egyptian Empire. The pyramids, erected nearly 4,000 years ago during the Fourth Dynasty, were intended as monuments to god-kings. Even today, they remain the largest cut stone edifices on the globe. They testify to the great organizational capabilities of ancient Egyptian civilization, and also — quite unintentionally — to the changes wrought by Egyptian civilization on the river valley that was its livelihood.

The Nile Valley was wooded and even marshy in many parts in the early years of the Egyptian Empire. Under the increasingly centralized, theocratic Egyptian state, however, the trees were felled and the marshes drained. Artificial irrigation and drainage combined with the natural blessings

▶ Sheep graze a once-forested Turkish hillside. In the Mediterranean as well as the Sahel, overgrazing has stripped the land of its original vegetation.

▼ A refugee at Nouakchott carries a load of scarce firewood. Wood is the chief fuel throughout the Sahel, and the role of trees in stabilizing dunes and ameliorating climatic extremes often gives way to the demands of the hearth.

▲ Bas relief panels in one of the tombs at Giza show agriculture along the Nile. Lines of workers wield scythes or carry sheaves of grain, probably barley or a form of wheat.

▶▶ The Sphinx at Giza is probably a monument to Cephren.

of the Sun and soil to make Egypt an exceptionally productive agricultural region.

The intensive agriculture that filled the granaries also contributed to the spread of the desert into the valley of the Nile. By cutting trees and eliminating the native shrub cover, Egyptian farmers opened the way to wind and water erosion. When the pyramids were built, they looked out over a productive landscape. Historical records describe

Giza as a rich, "black land," as opposed to the "red land" of the desert beyond.

The majestic modern setting of the pyramids in a sea of sand is as much a testament to the lasting damage the first great civilization inflicted on its homeland as it is to the power that was Egypt. Although the rich, renewing water of the Nile helped ancient Egyptian civilization remain vital for more than a hundred generations, eventually

▲ The burial pyramid of Cheops is the oldest and largest at Giza.

▶ The pyramids at Giza were built 4,500 years ago in a civilization based on the wealth of the Nile. These two commemorate Cheops, at right, and his son Cephren.

the rulers of the Nile proved unable to escape the Ozymandian fate of antiquity.

Many other civilizations have followed ancient Egypt along the shores of the Mediterranean. Here, among blue waters and sun-drenched islands, the Greek and Roman Empires flourished. Like the Egyptians, they left behind the ruins of great achievements and another legacy of desertification and land exhaustion. One place where this is particularly evident is the ruin of Ephesus, an ancient Greek city in what is now Turkey.

At its zenith 2,400 years ago, the wealth of Ephesus was proverbial. Its Temple of Artemis was

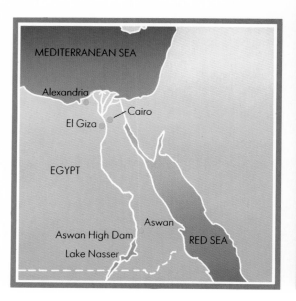

considered one of the Seven Wonders of the World. A center of learning, Ephesus boasted libraries and medical instruction that drew scholars from cities all around the Aegean. It was at Ephesus that the Roman Emperor Julian studied with the famous Sophist and magician, Maximus.

PRESSURE OF CIVILIZATION

In the end, Ephesus was destroyed by its own success. The ruins of edifices like the 24,000-seat amphitheater reflect the glory of ancient Ephesus, but the pattern of environmental alteration that doomed the city can best be seen under a microscope. The changing vegetation in the area has left a record in the form of pollen. Each species

▶ The library at Ephesus.

▼ The hills around the ancient city of Ephesus once were covered with oak.

of flowering plant produces a specific configuration of pollen. Because they resist deterioration, pollen grains can be recovered from dated sedimentary strata and used to determine the flora of earlier times.

At Ephesus, the predominant pollen type from 4,000 years ago is oak. These slow-growing, deep-rooted trees were the native vegetation at the time the area was first settled. Their acorns may have been a staple until other forms of agriculture took root. In later layers, the oak pollen gives way to that of plantain. A common weed in today's lawns and pastures, plantain colonizes cleared land. Its presence indicates the transformation from forest to grazing lands for sheep and goats.

Wheat pollen predominates in soil samples from a period 2,000 years ago, when Ephesus was at its zenith. With greater yields per acre than either acorns or livestock, wheat helped the citizens of Ephesus move their culture far beyond subsistence to create some of the glories of their age. The transformation from forest to wheat fields over 2,000 years made the land more productive, but it ultimately produced the ruin of the city.

The fatal mistake at Ephesus was the interruption of the water cycle. Before the arrival of settlers, the seasonal rains fell on wooded slopes. Some of this moisture quickly turned to mist and evaporated and some was carried away by mountain streams. A great deal, however, was retained by the deep roots of the oaks. Slowly, over many months, this stored moisture was released as transpiration, moderating the difference between the wet and dry seasons.

When the forest was cleared for pasture, the water equation changed. The grassland could not retain as much of the rain, and more water was released immediately as run-off, eroding ever deeper gullies in the vulnerable soil. This process accelerated when pastures were plowed and planted to crops like wheat. Many of the fields were wasted by soil erosion, so that barren hills eventually replaced what had been forested slopes.

POLLEN ANALYSIS

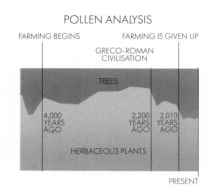

▲ The relative abundance of plant pollens is one indicator of land use around Ephesus. The proportion of tree pollens relative to grasses and farm crops fell as the city rose to prominence in the ancient world.

◄ The amphitheater at Ephesus, which seated 24,000.

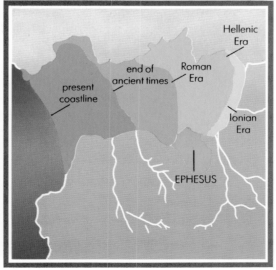

223

▶ Ruins of Ephesus. The grove of trees at right gives an idea what the earlier landscape may have been like.

As the centuries passed, the effects of deforestation, grazing, and farming began to be felt in the harbor that was the city's lifeline. Erosion from the surrounding hills was filling the harbor with silt. At least four times the city had to be moved closer to the retreating mouth of the bay. Despite these efforts and considerable dredging, the harbor was too shallow by the ninth century to accommodate the Byzantine fleet. Eventually the bay was almost entirely filled in, and Ephesus declined into obscurity.

Land that has been deforested does not easily recover. As long as a frontier remains, it is easier simply to move on. As the countryside around the Mediterranean was depleted, the eyes of empire fixed on undeveloped Europe. Initially spread by conquest, Mediterranean culture took root with religious conversion. Christianity proved particularly suited to both the spiritual and temporal needs of northern Europeans at the dawn of their era of world supremacy. In a seemingly boundless, unexploited continent, the new religion sanctioned human dominion over the barbarous wilderness of beech, oak, and spruce.

HIGH COSTS OF DEVELOPMENT

Certain indigenous European religions, such as that of the Druids, believed the trees were sacred. Therefore Christians sometimes made a point of cutting down the great trees as a symbol of the destruction of pagan worship, After the fall of Rome, thousands of new Christian communities were established in the forests of Europe. Often a small church would be built in the forest. Then more trees would be cut to enable people to begin farming. Over the centuries, the village would become a town, and perhaps even a city.

Paris is one example. The Cathedral of Notre Dame stands on an island in the Seine River where early Christian settlers built a church in the sixth century. The great cathedral was built 600 years later, as the city of Paris began to grow up around it. Today Notre Dame is surrounded by a great

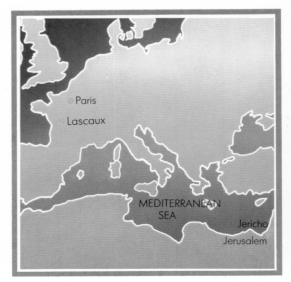

ROLES OF OZONE

As far as life on Earth is concerned, the place for ozone is in the stratosphere. At 40,000 feet, this unstable trio of oxygen atoms absorbs ultraviolet radiation. Near the ground, the same chemical is a dangerous and increasingly common pollutant.

Even quite low concentrations of ozone cause lung damage under prolonged exposure, and the high levels common in many American cities are a definite health hazard. Ozone is equally harmful to plants, destroying chlorophyll and lowering the plant's resistance to diseases and pests. Crop damage in the United States due to ozone may total over $5 billion a year.

Until the Industrial Revolution, the main producer of lower-level ozone was lightning. The brisk, charged-up smell of air after a thunderstorm is one of its signatures. Its rapid accumulation in the lower atmosphere during the last 20 years is due to the reaction of sunlight with the ubiquitous nitrogen oxides and hydrocarbons produced primarily by cars and factories. Even trees may play a part in its production. William Chameides of the Georgia Institute of Technology told *Science News* magazine in 1988 that his research indicates that trees in urban areas are a major source of hydrocarbons. Harmless in themselves, the hydrocarbons produced by trees are nevertheless highly reactive. When they react with the nitrogen oxides produced by the burning of fossil fuels, they create more ozone. Since trees are central to other crucial chemical cycles, logging is no solution. A society that wants clean air must curtail the use of fossil fuels.

The depletion of ozone in the upper atmosphere also leads to its build-up below. The ultraviolet radiation that reaches the troposphere converts nitrogen oxides into ozone. Unfortunately, ozone in the troposphere does not rise to augment the layer in the stratosphere. The reactive nature that allows ozone to absorb ultraviolet radiation makes it too unstable to pass through the atmosphere unchanged.

While highly reactive compounds like hydrocarbons and nitrogen oxides have long been recognized as contributors to pollution, the damage to the stratospheric ozone layer comes from chemicals that were thought to be benign.

Chlorofluorocarbons, called CFCs, are members of a class of inert compounds commonly used as refrigerants, aerosol propellants, and solvents. Because they do not react readily, CFCs are long-lived. A CFC molecule persists in the atmosphere for an average of 100 years. Eventually they rise into the stratosphere, where intense ultraviolet radiation destroys their chemical bonds, releasing chlorine. The chlorine in turn splits ozone into molecules of oxygen and chlorine monoxide. The presence of chlorine monoxide is an indicator for ozone destruction.

For life on the continent below, depletion of the ozone layer is a very serious matter. Ever since free oxygen first accumulated in the atmosphere, the ozone layer has blocked most ultraviolet radiation from reaching the surface of the Earth and made terrestrial life possible. More ultraviolet radiation means more skin cancer for land dwellers like ourselves, more cataracts, more immune system problems, more mutations, and ultimately — if it becomes intense enough — extinction.

Predictions made more than 10 years ago based on then-current amounts of chlorofluorocarbons estimated serious drops in ozone levels over the next 50 to 60 years. Since then the production of CFCs has increased despite bans on their use as aerosols, and the dramatic holes seen in the ozone layer over Antarctica have led many scientists to speed up the timetable. The alarming statistics have also inspired an intense effort to

▶ Los Angeles freeway. Exhaust from vehicles mingles with other hydrocarbons in urban air to become by far the largest source of low level ozone.

▶▶ For cities like Los Angeles, whose dependency on the automobile is nearly total, a solution to the dangers of air pollution seems no closer after years of study and expense.

find out why the seasonal holes have outstripped expectations.

The ozone hole appears in the Antarctic spring and fills in during the fall. The seasonal cycle is probably a function of the extreme polar climate. Frozen chlorine compounds may be trapped in the polar stratospheric clouds during the Antarctic winter, then released to do their destructive work when the Sun returns.

Joe C. Farman of the British Antarctic Survey — the scientist who first described the ozone hole — thinks that CFCs rise into the stratosphere in equatorial latitudes and that the chlorine compounds resulting from their breakdown are then sucked down into the polar vortex as it cools. More recent research by Margaret Tolbert and co-workers at SRI International in Menlo Park, California, indicates that the action of chlorine molecules may be speeded by the presence of sulfuric acid.

Previously implicated in episodes of global cooling following volcanic eruptions, the mixture of sulfuric acid and water is also produced by biological processes and by the burning of fossil fuels. A combination of CFCs and sulfuric acid could explain why the ozone layer has diminished so much faster than expected.

Despite the urgency of the quest, confirmation and measurement of ozone destruction has been difficult. Stratospheric ozone is rare even when it is most abundant, and only since the mid-1970s have detectors been able to record its fluctuations reliably. The detectors also must get to where the

ozone layer is suffering its greatest depletion, a journey made difficult and dangerous by the extreme weather conditions at the poles.

In August and September 1987, a modified U2 spy plane made several flights into the Antarctic stratosphere. The flights confirmed the seasonal hole in the ozone layer, and measurements of chlorine monoxide over 500 times normal levels established CFCs as the culprit. On September 17, more than 30 countries signed an international protocol calling for a freeze and eventual 50 percent reduction in CFC production in developed countries.

Many scientists think the compact does not go nearly far enough. Studies made before the conference by the United States Environmental Protection Agency concluded that a 95 percent worldwide cut in CFC production was needed. Since CFCs are already present in huge quantities in refrigerator systems, air conditioners, foam insulation, and other products around the world, as well as in the atmosphere, the ozone layer would probably continue to decline for several years even if all new production stopped tomorrow.

Still, the very instability of ozone means that its destruction is reversible. Like the anaerobic organisms that first faced atmospheric oxygen more than three billion years ago, humans may be reaching an environmental crossroads caused by their own activities. Unlike our unicellular forebears, we are able to contemplate our own dilemma, and perhaps to find a solution.

metropolis. Viewed from the top of the Eiffel Tower, Paris spreads out in all directions to the horizon.

Berlin presents a similar picture, with one important exception. Like Paris, it began as a village in what was once a dense forest, and like Paris most of its surrounding forest was logged by the end of the eighteenth century. But today satellite images of Berlin show dark green patches scattered through the surrounding countryside. New trees are growing to replace some of those cut.

Several generations of trees have been planted and harvested since the Romans first moved into northern Europe. Taken as a whole, however, European civilization has produced the same sort of deforestation previously seen in Egypt, Greece, and Rome. It survived and prospered by extending

its resource base to truly global proportions, drawing on Asia, Africa, and the Americas to provide the native bounty that was no longer available at home.

Now, less than 500 years after Christopher Columbus first laid eyes on the New World, America too has been stripped of its virgin forests and subjected to tremendous physical abuse. The twentieth century, in particular, can be read as a litany of man-made environmental disasters, from the Dust Bowl to the death of the Great Lakes to Love Canal.

America has not been toppled by these calamities, however. This is partly due to its youth and the great magnitude of its natural wealth, and partly because in areas like the Great Lakes, real improvements have been made in the environment. However, the greatest reason for the continued preeminence of America specifically and the industrial nations in general has been their ability to tap new energy sources.

It is estimated that a person in a modern, developed nation consumes as much energy in a day as a person living 1,000 years ago would use in a year. Nowhere is the tremendous energy consumption of modern industrial society more evident than over the United States on a clear night. Viewed from space, a map of the United States is traced in light from Miami to Seattle. Some of this electricity comes from nuclear plants and hydropower, and a few million Americans heat their homes with wood or sunlight, but the lavish abundance of American society comes from petroleum and coal.

DEFORESTATION

The price is high. In some areas of the American Midwest, an estimated one and a half bushels of topsoil is lost for every bushel of corn produced under modern farming conditions. Production has been kept up through the heavy use of artificial fertilizers, most of which are petroleum-based, and

▼ Logging continues to be a major industry throughout the world. In Europe, all the remaining forests are the result of planning and reforestation. However, acid rain and ozone damage now are taking their toll on the successors to the wilderness the Romans encountered. Photograph: Steve Vidler, Australian Picture Library.

◀ Crowded, vibrant, and almost wholly unnatural, lower Manhattan is the apothesis of mind over nature. New York cannot grow its food, supply its own power, or dispose of its garbage.

▼ Nördlingen, West Germany. Founded more than a thousand years ago, this Bavarian manufacturing town is still centered within the walls built between the fourteenth and sixteenth century.

◀ Corn harvest in Grand Island, Nebraska.

◀◀A Nebraska wheatfield. Until the arrival of farmers, the Great Plains were covered with a multitude of grasses, some of which grew tall enough to hide a mounted horseman. They have been replaced with a few varieties of farm crops, a transformation both productive and dangerously vulnerable. Photo: Australian Picture Library.

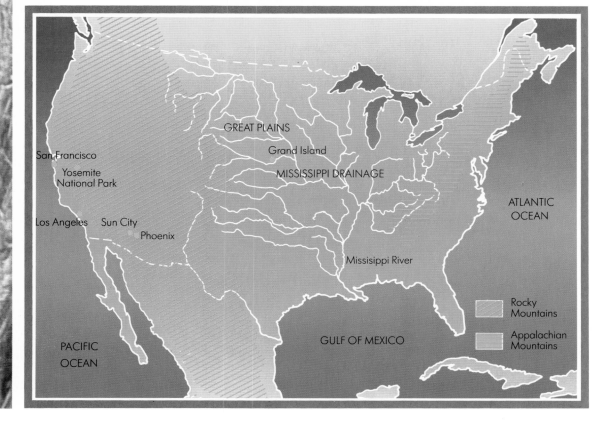

the huge machinery that has replaced most hand labor on the American farm. As natural fertility declines, it takes still more energy to produce the same crop. The problem is one of diminishing supply. Properly tended, a forest planted today is a resource for our children. But the coal now being formed in swamps and taiga will not be ready for millennia. Unless new supplies are found, the expense of extracting fossil fuels will soon become prohibitive.

Even if the supply were infinite, the large-scale consumption of coal and oil would be problematic. In effect it accelerates one aspect of the carbon cycle to the point where the network of systems that keep the atmosphere and climates stable is strained, perhaps to the breaking point. At Mauna Loa Marine Atmospheric observatory on the island of Hawaii, scientists have chronicled the changing nature of the Earth's atmosphere. Even here, far from industrial centers, the concentration of carbon dioxide in the atmosphere has been increasing steadily since observations began in 1958.

Carbon measurements taken from tree rings indicate that before the Industrial Revolution concentrations of carbon dioxide were between 260 and 280 parts per million, while now they are

▶ Grain elevators. Richly supplied with soil and technology, the United States raises more grain than it can consume.

▼ This harvester can cut eight rows of corn at a time.

FOSSIL FUELS

THE transformation of palm trees and dinosaur flesh into coal and oil is a concept that is known — if not really understood — by practically everyone over the age of ten.

How is this alchemy of energy accomplished? Like the karst landscapes of China, "fossil fuels" are part of the carbon cycle. In the biosphere they act as storehouses of hydrocarbons, releasing the gas gradually and helping to keep atmospheric levels stable.

In human terms, however, these concentrated and transformed remains of living organisms are the key to the treasure house. They provide tremendous amounts of usable energy packaged in a form that can be transported, stored, and sold.

Fossil fuels begin with an abundance of life and a shortage of oxygen. In modern times, these conditions are fulfilled by swamps like the Okefenokee in Georgia, where the warm, shallow water does not contain enough oxygen to consume all the organic material that falls to the bottom. Over time, this partly decomposed matter forms a black substance called peat. In the Okefenokee, peat accumulates at the rate of about an inch every 50 years. Some peat bogs there are 16 feet thick, and elsewhere they reach a thickness of as much as 50 feet.

Coal forms when peat is sufficiently compressed by the weight of the growing deposits above to force out water and volatile compounds, leaving a concentrated mass of carbon. When partially decomposed plankton and other organic matter are overlaid with accumulating sediment, the resulting heat and pressure may yield an even more concentrated energy source, petroleum.

Three hundred million years ago during the Carboniferous period, much of the world resembled the Okefenokee. The land masses were grouped into larger continents than we have today and surrounded by shallow seas and marshes. What is now the interior of the United States was mostly under water. The climate was humid, and plant life was incredibly luxuriant. Hundred-foot-tall ferns, giant horsetails, club mosses, and primitive conifer trees grew thick along what is now the east coast of North America. Today, the record of this lush growth lies in the anthracite coal fields of Appalachia, Wales, and England, and the Saar Basin of continental Europe.

Tremendous reserves of coal and oil still remain underground, but their cost is rising steadily both in dollar and environmental terms. The easily accessible reserves have been exploited. New ones will be expensive to locate and expensive to work. Furthermore, the dangers of air pollution and global warming — to which fossil fuels are a major contributor — are becoming ever more evident.

Since the Industrial Revolution, the rate of consumption of fossil fuels has vastly exceeded their rate of creation. The formation of coal and oil takes millions of years and special geologic circumstances. Environmentalists have long pointed out that even if society chooses to ignore the environmental implications of unrestricted use of fossil energy, the supplies

▶ The Okefenokee Swamp in south Georgia and Florida bears comparison with the forests of the Carboniferous period in which most of the world's coal was laid down. Lignite coal, which was formed as recently as the Tertiary period less than 50 million years ago, comes from an environment almost identical to the Okefenokee.

▶▶ Using a tripod for leverage in the shaky soil, geologists take core samples in a peat bog in the Okefenokee.

cannot last forever. Their opponents have disagreed about when, not whether, fossil fuels will become too scarce to count on.

Recently, however, this venerable truism has come under attack by Thomas Gold, the same astrophysicist who suggested that our minds run backward as the universe rewinds.

Thomas Gold suggests that in addition to the fuels created from biological leavings, large quantities of hydrocarbons that had arrived early in the planet's history were transformed into gas by the heat and pressure 100 miles below the Earth's surface. As this gas migrated upward, like magma along fissure lines, some of it was converted to oil.

Gold does not contend that all petroleum has extraterrestrial sources. Most is of biologic origin. But as the identified sources become depleted, he thinks it is worthwhile to test his theory by drilling deeper than most equipment can go. One such test was done north of Stockholm in 1987. Drillers found black sticky matter 20,000 feet below the surface in an area where there was no fossil evidence to indicate hydrocarbons.

Most analysts think the sludge is simply oil from the drilling operation itself. Gold thinks it is treasure from the deep. Even if he is right, however, pollution problems may make his discovery a mixed blessing.

▲ Dr. Syukuru Manabe of Princeton University has developed climate models that respond to changes in carbon dioxide percentages in the atmosphere.

In 1983, *Science* magazine reported that an estimated 1.1 tons of carbon as carbon dioxide is released every year for every person on Earth. In the United States, the per capita release is five tons per year.

Rising carbon dioxide levels are linked to increasing temperatures, since carbon dioxide traps more of the Sun's energy than does oxygen. Many scientists, including James Hansen of the NASA Goddard Institute of Space Studies, believe that the devastating drought that hit the United States in the summer of 1988 was a harbinger of life under a "greenhouse" of carbon dioxide and methane gases.

Professor Syukuru Manabe of the Geophysical Fluid Dynamics Laboratory at Princeton, New Jersey, has created a computer model that attempts to show how Earth's climate would respond to increasing levels of atmospheric carbon dioxide and other greenhouse gases. According to projections by some atmospheric chemists, levels of carbon dioxide might reach as high as 600 parts per million by the year 2050. If Manabe's model accurately portrays Earth's climate, that change would raise the average temperature of the Earth between 4 and 8 degrees Fahrenheit.

Even a 4-degree rise in temperature could have profound effects, especially on the distribution of rainfall. In Manabe's model, increased surface evaporation leads to greater dehydration in inland areas. It appears the greatest impact would be on the great granary regions of the American Midwest, where soil moisture content could be reduced by 50 percent. The same process would be felt to varying degrees across the planet. Most regions in

about to reach 350 parts per million. Human actions almost certainly are largely responsible for the changing ratio. Every time a tree is cut and burned and not replaced, a little more carbon dioxide is released into the atmosphere and the ability to absorb the gas is slightly diminished. Burning coal and oil releases in a rush the carbon dioxide that was taken up bit by bit in earlier eras.

▶ Lignite coal, the softest and most recently formed of coal deposits, is accessible through open pit mining.

the northern hemisphere, except for India and Southeast Asia, would suffer drought. What are now prime agricultural districts in China, the Soviet Union, and Europe would be much less productive.

The same warming and evaporation that may bring drought to many interior regions has the potential to drown hundreds of coastal communities and bring increased rainfall to thousands more. Computer projections done in 1987 at the University of East Anglia predict that the greenhouse effect will cause a 2- to 3-inch rise in the average sea level over the next 40 years as ice sheets begin to melt. Such a rise would significantly increase shoreline erosion and extend salt pollution into freshwater lakes and rivers. If the trend continues, many coastal cities would be uninhabitable by the end of the next century.

Researchers at the National Oceanic and Atmospheric Administration report that the number of giant icebergs breaking off the Ross Ice Shelf in Antarctica has increased substantially since 1986. One enormous chunk, 750 feet thick and covering twice the area of the state of Rhode Island, has

been floating off the coast of Antarctica since October 1987.

Other results of rising carbon dioxide levels are less obvious. River levels might rise, an unexpected effect in a drier climate. The reason, according to an article in a November 1984 edition of *Nature*, is that plants adapt to increased carbon dioxide ratios by tightening their pores to conserve water. When plants need less water, run-off increases and rivers rise. Other research indicates that insects eat more when feeding on leaves with a high carbon dioxide level. These responses point out one of the imponderables of any long-term change in climate: we do not know how adaptable individual species are until they have been put to the test.

At the same time more carbon dioxide is being released into the atmosphere, the Earth's ability to absorb carbon dioxide is being compromised. The major repository for the gas is water, and the oceans' absorption capacity is tremendous but not infinite. Coral reefs and coastal wetland flora take up carbon dioxide from the water, making room for more, while forests serve the same function on

▲ An open pit mine in Illinois. Much of the coal mined in the United States is burned to produce electrical power, although gleaners in some coal-rich areas pry it out of road cuts and ditches, and take it home for their stoves.

237

OTHER WORLDS

THE rhythms of Earth are slow, and humans are impatient. Few of us want to wait 100 years — let alone several thousand — to learn the real consequences of ozone depletion or global warming. We want predictions we can act on now, and scientists have not been slow to supply them.

Like the old adage, "red sky at night, sailor's delight," modern forecasts are based on taking data from the past and projecting them onto our understanding of the way Earth's systems work. At the basis of every prediction is the assumption that the cycles of the future will follow the rules we have observed from the past.

In modern laboratories, the premier predictive tool is the computer model. Syukuru Manabe's projections of dry years in the world's agricultural heartlands come from plugging higher carbon dioxide levels into mathematical calculations of evaporation rates, jet streams, temperature variations, cloud cover, ocean currents, and dozens of other processes.

The mathematics of such an enterprise are complex and laborious. A worldwide climate model can involve hundreds of billions of computations and still not approach the intricacy of the real thing. The result of all this specificity, moreover, is likely to be a very general conclusion: "All we can say is that maybe the mid-continental United States becomes dryer," says Dr. Manabe.

The more precise a prediction attempts to be, the more likely it is that its inevitable missing pieces will render it useless. Manabe's model does not include plants, and so misses an important element in the transport of water. Climate models from the early 1980s failed to predict the ozone hole over the Antarctic because they left out the effects — which were not fully understood — of polar ice clouds. The planet is infinitely complex, and no model can be expected to duplicate it. However, the computers do allow researchers to test their hypotheses promptly.

The best way to evaluate a computer model is to compare it with reality. Climate predictions for the future can be compared with the model's performance using known data from the past. If the computer prediction matches what actually happens, the model gains credibility. Another possibility is to build a "real" model to test the mathematical one. Laboratory scientists do this when they repeat experiments in an attempt to replicate their findings.

A particularly ambitious experiment in model making is getting under way in 1989, as eight volunteers are sealed inside a 2.5-acre mini-world called Biosphere II. Located in Oracle, Arizona, on the outskirts of Tucson, Biosphere II is an attempt to create a self-sustaining world under glass. For two years, if the experiment is successful, the inhabitants will have only electronic communication with the outside world. All their food, and all their weather, will be produced inside the complex. The mini-world has its own ocean — with a coral reef nourished by mechanically generated waves — and its own rain forest and desert. Vegetables are grown hydroponically, using water recycled from the ponds where tilapia, a fast-growing food fish, are raised.

Because the system is closed, no agricultural poisons or strong sterilizers can be tolerated. Biological controls including ladybugs and predatory wasps will be used to control pests.

Waste water will be passed through filters and grazed by microorganisms before being supplied to the chickens. In the absence of a chilly upper troposphere, rain clouds will be created by cooling coils above the tropical forest.

Biosphere II cannot approach in scale and complexity the operations of Biosphere I — the Earth itself. For one thing there isn't time. Although the project will depend on the natural cycles

of flora and fauna to keep carbon dioxide and oxygen in balance, those make up only a fraction of the carbon cycle in the world outside. About 80 percent of the carbon exchanged between ground and air depends on erosion and operates on a scale of about 500,000 years. Neither can the air under an 80-foot ceiling duplicate the wind systems and chemical diffusions of 400 miles of atmosphere.

Biosphere II is a project of Space Biospheres Ventures, a private company which plans to license the technology to governments planning space colonies. Its greatest utility, however, will probably be for those of us who prefer to stay at home. All the systems will be monitored carefully for the duration of the experiment, and the data no doubt will find its way into upcoming models of the uncovered world.

land. When these carbon dioxide "sinks" are destroyed, their loss exacerbates problems created by the profligate consumption of fossil fuel. Furthermore, the ocean's capacity for absorbing and storing heat has served to mask the amount of warming that already has taken place. V. Ramanathan, an atmospheric scientist at the University of Chicago, estimates that greenhouse heat already stored in the ocean will raise global temperatures 2 to 5 degrees Fahrenheit over the next 30 to 50 years.

The living systems that comprise the best defense against dangerous warming are not adequately protected. One-third of the world's equatorial rain forests are contained within the Amazon Basin, and it is here the contemporary assault on the rain forests has been most intense. Thousands of square miles a year are currently being logged. Under the tropical Sun and rain, the shallow soils are quickly eroded. The land loses its capacity to hold moisture, and at the same time the weather has changed toward drier conditions, much as occurred around Ephesus more than 1,000 years ago.

Another result of overcutting tropical forests is damage to the coral reefs. The removal of the vegetation leads to massive soil erosion. Dumped by rivers and ocean currents onto the reef communities, the sediment coats the coral with a film that can kill in a matter of days. Sediment due to logging, agriculture, and land clearing has already damaged reefs in Australia, Puerto Rico, and Hawaii. Thus the effect of human actions runs through the ecosystem, rippling from top to bottom.

For the first 5,000 years of human civilization, people were dependent on simple, easily accessible energy sources such as wood, sunlight, and flowing water. Contained within the finite limits of the energy and environment available to them, early cultures repeatedly played out the same pattern of rise and decline.

Modern civilization, however, has been able to extend its life by making use of ancient, stored forms of energy. The one thing the alchemy of fossil fuel energy has not been able to overcome is civilization's age-old tendency toward resource depletion and environmental degradation. In fact, we are consuming more resources today than ever before, and the effects of our actions are felt all over the world.

Today, our species dreams of totally controlling its environment. The impulse is evident everywhere from self-contained retirement communities like Sun City, Arizona, to Biosphere II. Human beings have made tremendous strides into space during this century, but the experience of floating free of the globe, so long envisioned in science fiction, has actually proved our frailty against the cosmos.

Space may be the last frontier, but it is hardly a welcoming place. In practical terms, it is a vast and hostile realm, where the slightest technical failure means death. For better or for worse, the only welcoming world we know is this one.

► Sun City, Arizona. The philosophical opposite of Biosphere II (described on pages 238-39), the retirement community is nearly totally dependent on the outside. Heavy development in southern Arizona is leading to a crisis as the water table recedes to dangerous levels.

FURTHER READING

Two informative publications geared to the nonprofessional are *Science News* (Science Service), a weekly, and *The Sciences* (New York Academy of Sciences), a monthly. Two other valuable magazines, *Science* and *Scientific American,* are more technical in their content and presentation.

THE THIRD PLANET

Angelo, Joseph A. Jr. 1985. *The Extra Terrestrial Encyclopedia: Our Search for Life in Outer Space.* New York: Facts on File Publications.
Straightforward reference work on exobiology.

Bartusiak, Marcia. 1986. *Thursday's Universe.* New York: Times Books.
Excellent introduction to prevailing theories of the formation and workings of the universe by a well-known science writer.

Browne, Malcolm W. "Debate Over Dinosaur Extinctions Takes Unusually Rancorous Turn." *New York Times.* January 14, 1988.

Cooper, Henry S. F. Jr. 1980. *The Search for Life on Mars: Evolution of an Idea.* New York: Holt, Rinehart & Winston.
Behind the scenes with the biological research team of the Viking I and II space probe.

Field, George B. and Eric J. Chaisson. 1985. *The Invisible Universe.* Boston: Birkhauser.
Astronomy outside the visible spectrum. Includes section on upcoming projects.

Goldsmith, Donald. 1985. *Nemesis: The Death-Star and Other Theories of Mass Extinction.* New York: Walker & Co.

Greenstein, George. 1988. *The Symbiotic Universe.* New York: William Morrow.
Greenstein, an astrophysicist, uses quantum theory to defend his contention that life is an essential component of the universe.

Hoyt, William Graves. 1976. *Lowell and Mars.* Tucson: University of Arizona Press.

LeMaire, T. R. 1980. *Stones from the Stars: The Unsolved Mysteries of Meteorites.* Englewood Cliffs, N. J.: Prentice-Hall.
Accessible treatment of meteorite theories and discoveries.

Morris, Richard. 1982. *The Fate of the Universe.* New York: Playboy Press.
Makes the case for both an open and shut universe. Morris, a physicist and poet, writes lucidly about extreme forces and vast expanses of time.

Raup, David M. 1986. *The Nemesis Affair.* New York: W. W. Norton & Co.
A prominent exponent of periodic extinction describes the history of the theory and his own evolution from critic to supporter. A fascinating insight into how science works.

Ronan, Colin A. 1982. *Science: Its History and Development Among the Earth's Peoples.* New York: Facts on File Publications.
A well-illustrated survey that shows the links between scientific theory and developing technology.

Sagan, Carl. 1980. *Cosmos.* New York: Random House.
A spur to the imagination by the best-known proponent of space science.

Tucker, Wallace, and Riccardo Giacconi. 1985. *The X-Ray Universe.* Cambridge: Harvard University Press.
History and applications of X-ray telescopes.

▶ Dust from the Sahara coats the sea floor and colors sunsets as far away as Miami.

▶▶ Coulees and scablands of the Columbia Plateau show the erosive power of water. Photograph: Harald Sund.

Following on page 245: Ice sheets and subsequent flooding stripped the soil away from the Columbia Plateau creating channels in the lava. Photograph: Harald Sund.

The Heat Within

Ballard, Robert D. 1983. *Exploring Our Living Planet.* Washington, D.C.: National Geographic Society.
Links the human history of the Andes and Cyprus with the geologic processes that brought mineral wealth to both regions.

Fisher, Arthur. "What Flips Earth's Field?" *Popular Science,* January 1988. 71-73.
Thermoremanent magnetism and its possible relation to climate and mass extinctions.

McPhee, John. 1981. *Basin and Range.* New York: Farrar, Straus, Giroux.
With the American landscape as his focus, McPhee shows how field geologists work, and how the theories of plate tectonics have affected their practice.

McPhee, John. 1983. *In Suspect Terrain.* New York: Farrar, Straus, Giroux.
More explorations of the American landscape, in the company of a critic of tectonic theory.

McPhee, John. "The Control of Nature (Volcano)."
The New Yorker. February 22 and 29, 1988.
Eruptions at Kilauea, Hawaii, and Heimaey, Iceland and their relation to plate tectonics.

Sullivan, Walter. 1984. *Landprints: On the Magnificent American Landscape.* New York: Times Books.
Geological and manmade features of North America. Includes a section on the channeled scablands of Eastern Washington.

Takeuchi, H., S. Uyeda and H. Kanamori. 1970. *Debate about the Earth: Approaches to Geophysics through Continental Drift.* San Francisco: Freeman Cooper & Co.
A lucid, textbook discussion of plate tectonics and paleogeomagnetism.

Life From The Sea

Cairns-Smith, A. G. "The First Organisms." *Scientific American.* June 1985. 90-92.

Cann, Rebecca L. "In Search of Eve." *The Sciences.* September-October 1987. 30-37.
The use of mitochondrial DNA to trace a common human ancestor, explained by one of the geneticists involved.

Cloud, Preston. "The Biosphere." *Scientific American.* September 1983. 176-182.

Hartman, H., J. Lawless and P. Morrison, eds. 1987. *Search for the Universal Ancestor: The Origins of Life.* Palo Alto, Calif.: Blackwell Scientific Publications.
Fossil and genetic evidence of early life forms.

Kvenvolden, Keith. A., ed. 1974. *Geochemistry and the*

Origin of Life. Stroudsburg, Pa.: Dowden, Hutchinson & Ross.

Includes some of the seminal papers on the chemical origins of life.

Lovelock, James. 1979. *Gaia: A New Look at Life on Earth.* Oxford University Press.

Arguments for the theory of Earth as a living entity. Includes an explanation of cybernetics.

Margulis, Lynn. 1982. *Early Life.* Boston: Science Books International.

Easy-to-follow account of Earth's environment before and after the advent of life, by an associate of James Lovelock. Explains the effects of oxygen on cell biology.

Newell, Norman D. "The Evolution of Reefs." *Scientific American.* June 1972. 50-65.

Drawing their substance from the ocean around them, reefs record in their structure environmental changes spanning millennia.

Schopf, J. W., ed. 1983. *The Earth's Earliest Biosphere.* Princeton University Press.

Walker, James C. G. 1977. *Evolution of the Atmosphere.* New York: Macmillan.

Weiss, R. "Seekers of the Ancestral Cell Debate New Data." *Science News.* February 2, 1988. 36.

PATTERNS IN THE AIR

Cloudsley-Thompson, J. L., ed. 1977. *Man and the Biology of Arid Zones.* Baltimore: University Park Press.

Cloudsley-Thompson, J. L., ed. 1984. *Sahara Desert.* New York: Pergamon.

Ellis, W.S. "The Sterilization of the Sahel." *National Geographic.* August 1987. 146-171.

Human impact on the fragile ecology of the sub-Sahara.

Frakes, L. A. 1979. *Climates Throughout Geologic Time.* Amsterdam: Elsevier Scientific Publishing Company.

Ties continental movement to changes in climate, and explains the relationship of climate to composition of the atmosphere.

Ingersoll, Andrew P. "The Atmosphere." *Scientific American.* September 1983. 162-174.

Focuses on the role of the atmosphere in distributing solar radiation.

Kasting, James F., Owen B. Toon and James B. Pollack. "How Climate Evolved on the Terrestrial Planets." *Scientific American.* February 1988. 90-97.

Lockwood, J. G. 1979. *Causes of Climate.* New York: Wiley.

Newell, Norman D. "The Evolution of Reefs." *Scientific American.* June 1972. 50-65.

Swift, Jeremy. 1975. *The Sahara.* New York: Time-Life Books.

RIDDLES OF SAND AND ICE

Bailey, Ronald H. 1982. *Glacier.* Alexandria, Va.: Time-Life Books.

Books in the Time-Life series have excellent photographs and biographical sketches.

Chorlton, Windsor. 1983. *Ice Ages.* Alexandria, Va.: Time-Life Books.

Halacy, D. S., Jr., 1978. *Ice or Fire? Surviving Climatic Change.* New York: Harper & Row.

Includes futuristic proposals for altering Earth's climate.

Lurie, Edward. 1960. *Louis Agassiz: A Life in Science.* The University of Chicago Press.

Monastersky, Richard. "Shrinking Ice May Mean Warmer Earth." *Science News.* October 8, 1988. 230.

Schultz, Gwen. 1963. *Glaciers and the Ice Age: Earth and Its Inhabitants During the Pleistocene.* New York: Holt, Rinehart & Winston.

Simple introduction to glaciology.

Stammel, Henry, and Elizabeth Stammel. 1983. *Volcano Weather: The Story of 1816, the Year Without a Summer.* Newport, R.I.: Seven Seas Press.

Entertaining historical sleuthing into the consequences of the eruption of Mount Tambora.

THE HOME PLANET

Bach, Wilfred. 1984. *Our Threatened Climate: Ways of Averting the CO_2 Problem Through Rational Energy Use.* Dordect, Holland: Reidel.

Reviews international studies on CO_2 emissions and energy use.

Foss, C. 1979. *Ephesus After Antiquity: A Late Antique, Byzantine and Turkish City.* Cambridge University Press.

Freundlich, Naomi. "Biosphere." *Popular Science.* December 1986. 54-56.

Describes the Biosphere II project.

Johnson, Paul. 1978. *The Civilization of Ancient Egypt.* Ch. 1. New York: Atheneum.

Illustrates the scale of Egyptian agriculture and irrigation projects.

Kerr, Richard A. "Greenhouse Warming Still Coming." *Science.* May 2, 1988. 573-574.

Kerr, Richard A. "The Carbon Cycle and Planet Warming." *Science.* December 9, 1983. 1107-8.

A glimpse at the complexity of the carbon cycle and the greenhouse effect.

Lamb, H. H. 1982. *Climate, History and the Modern World.* New York: Methuen.

Lhote, Henri. "Oasis of Art in the Sahara." *National Geographic.* August 1987. 181-185.

More photographs of the rock paintings at Tassili N'ajjer.

Meiggs, Russell. 1982. *Trees and Timber in the Ancient Mediterranean.* Oxford: Clarenden Press.

Explores long-term effects of land-use practices of early Mediterranean cultures.

Monastersky, Richard, "Clouds Without a Silver Lining." *Science News.* October 15, 1988. 249-251.

The role of stratospheric clouds in ozone depletion.

Monastersky, Richard, "Decline of the CFC Empire." *Science News.* April 9, 1988. 234-236.

Industry responses to restrictions on chlorofluorocarbons.

Russell, Dick, with Russell King. "Politics of Ozone: Delay in the Face of Disaster." *In These Times.* August 17-30, 1988.

Political perspective on ozone depletion.

Starr, Douglas. "How to Protect the Ozone Layer." *National Wildlife.* December/January 1988. 26-28.

Thirgood, J. V. 1981. *Man and the Mediterranean Forest.* Ch. 1. New York: Academic Press.

BIBLIOGRAPHY

THE THIRD PLANET

Bartusiak, Marcia. 1986. *Thursday's Universe.* New York: Times Books.

Browne, Malcolm W. "Debate Over Dinosaur Extinctions Takes Unusually Rancorous Turn." *New York Times,* January 14, 1988. 19.

Cooper, Henry S. F. Jr. 1980. *The Search for Life on Mars: Evolution of an Idea.* New York: Holt, Rinehart & Winston.

Field, George B. and Eric J. Chaisson. 1985. *The Invisible Universe.* Boston: Birkhauser.

Goldsmith, Donald. 1985. *Nemesis: The Death-Star and Other Theories of Mass Extinction.* New York: Walker & Co.

Hartmann, William K. "Cratering in the Solar System." *Scientific American.* January 1977. 84-86.

"Hints of Planets Circling Nearby Stars". *Science News.* August 13, 1988. 103.

LeMaire, T.R. 1980. *Stones from the Stars.* Englewood Cliffs, N.J.: Prentice-Hall.

McAleer, Neil. 1982. *The Cosmic Mind-Boggling Book.* New York: Warner Books.

Monastersky, Richard. "K-T Mass Extinctions: Abrupt or What?" *Science News.* October 31, 1987. 277.

Morris, Richard. 1982. *The Fate of the Universe.* New York: Playboy Press.

Raup, David M. 1986. *The Nemesis Affair.* New York: W. W. Norton & Co.

Sagan, Carl. 1980. *Cosmos.* New York: Random House.

Tucker, Wallace, and Riccardo Giacconi. 1985. *The X-Ray Universe.* Cambridge: Harvard University Press.

Van Andel, Tjeerd H. 1985. *New Views on an Old Planet: Continental Drift and the History of the Earth.* Ch. 2. Cambridge University Press.

Wilford, John Noble. "Dinosaur Fossils Found in Alaska." *New York Times.* September 25, 1988.

THE HEAT WITHIN

Ballard, Robert D. and J. Frederick Grassle. "Strange World Without Sun." *National Geographic.* October 1977. 680-705.

Colbert, Edwin H. 1973. *Wandering Lands and Animals.* New York: E. P. Dutton & Co.

Glass, Billy P. 1982. *Introduction to Planetary Geology Vol. 2 Planetary Science Series.* Cambridge University Press.

Hamblin, W. Kenneth, ed. 1985. *The Earth's Dynamic Systems: A Textbook in Physical Geology.* Minneapolis, Minn.: Burgess Publishing Co.

Hekinian, Roger. "Undersea Volcanoes." *Scientific American.* July 1984. 46-55.

McPhee, John. 1981. *Basin and Range.* New York: Farrar, Straus, Giroux.

McPhee, John. 1983. *In Suspect Terrain.* New York: Farrar, Straus, Giroux.

McPhee, John. "The Control of Nature (Volcano)." *The New Yorker.* February 22 and 29, 1988.

Monastersky, Richard. "The Whole Earth Syndrome." *Science News.* June 11, 1988. 379.

Takeuchi. H., S. Uyeda and H. Kanamori, 1970. *Debate About the Earth: Approaches to Geophysics Through Continental Drift.* San Francisco: Freeman Cooper & Co.

Tapponier, Paul. "A Tale of Two Continents." *Natural History.* November 1986. 56-64.

Yulsman, Tom. "Plate Tectonics Revised." *Science Digest.* November 1985. 35.

Weisburd, Stefi. "Wildfires: Apocalypse Then and Now." *Science News.* October 12, 1985. 228.

LIFE FROM THE SEA

Bendall, D.S., ed. 1983. *Evolution from Molecules to Men: The Primary Lines of Descent and The Universal Ancestor.* Cambridge University Press.

Budyko, Mikhail I. 1986. *Evolution of the Biosphere.* Dordecht, Holland: Reidel.

Cairns-Smith, A.B. "The First Organisms." *Scientific American.* June 1985. 90-92.

Cloud, Preston. "The Biosphere." *Scientific American.* September 1983. 176-182.

Dorfman, Andrea. "Earth's Oldest Tenants." *Science Digest.* July 1984. 16.

Gribben, John. "Earth's Lucky Break." *Science Digest.* May 1983. 36-37.

Hartman, H., J. Lawless and P. Morrison, eds. 1987. *Search for the Universal Ancestor: The Origins of Life.* Ch. 3. Palo Alto, Calif. Blackwell Scientific Publications.

Kvenvolden, Keith. A., ed. 1974. *Geochemistry and the Origin of Life.* Stroudsburg, Pa: Dowden, Hutchinson & Ross.

Lovelock, James. 1979. *Gaia: A New Look at Life on Earth.* Oxford University Press.

Margulis, Lynn. 1982. *Early Life.* Boston: Science Books International.

Monastersky, Richard. "Vents Would Scald a Primordial Soup." *Science News.* August 20, 1988. 117.

Raloff, Janet. "Clues to Life's Cellular Origins." *Science News.* August 2, 1986. 71.

Schopf, J. W., ed. 1983. *The Earth's Earliest Biosphere.* Princeton University Press.

"The Search for Adam and Eve." *Newsweek.* January 11, 1988. 46-52.

Walker, James C.G. 1977. *Evolution of the Atmosphere.* New York: Macmillan.

Weisburd, Stefi. "The Microbes That Loved the Sun." *Science News.* February 15, 1986. 108-110.

Weiss, Rick. "Seekers of the Ancestral Cell Debate New Data." *Science News.* February 2, 1988. 36.

PATTERNS IN THE AIR

"Ancient Air Idea May Not Hold Water." *Science News.* August 27, 1988. 141.

Bradshaw, M. J., A. J. Abbott, and A. P. Gelsthorp. 1984. *The Earth's Changing Surface: An Introduction to Geomorphology.* Ch. 13. Oxford University Press.

Cloudsley-Thompson, J. L., 1977. *Man and the Biology of Arid Zones.* Baltimore, Md.: University Park Press.

Cloudsley-Thompson, J. L., 1984. *Sahara Desert.* New York: Pergamon.

Ellis, W. S. "The Sterilization of the Sahel." *National Geographic.* August 1987. 146-171.

Frakes, L. A. 1979. *Climates Throughout Geologic Time.* Amsterdam: Elsevier Scientific Publishing Company.

Hidore, John J. 1972. *A Geography of the Atmosphere.* 2nd ed. Dubuque, Iowa: Wm. C. Brown Company Publishers.

Ingersoll, Andrew P. "The Atmosphere." *Scientific American.* September 1983. 162-174.

Kasting, James F., Owen B. Toon and James B. Pollack. "How Climate Evolved on the Terrestrial Planets." *Scientific American.* February 1988. 90-97.

Lockwood, J. G. 1979. *Causes of Climate.* New York: Wiley.

Monastersky, Richard. "Amber Yields Samples of Ancient Air." *Science.* November 7, 1987.

Newell, Norman D. "The Evolution of Reefs." *Scientific American.* June 1972. 50-65.

Press, Frank, and Raymond Siever. 1982. *Earth.* New York: W. H. Freeman.

Pritchard, J. M. 1979. *Landform and Landscapes in Africa.* London: Edward Arnold.

Walker, James C. G. 1977. *Evolution of the Atmosphere.* New York: Macmillan.

RIDDLES OF SAND AND ICE

Bailey, Ronald H. 1982. *Glacier.* Alexandria, Va.: Time-Life Books.

Bower, B. "Shuttle Radar is Key to Sahara's Secrets." *Science News.* April 21, 1984. 244.

Chorlton, Windsor. 1983. *Ice Ages.* Alexandria, Va.: Time-Life Books.

Crowell, John C. 1982. *Climate in Earth's History.* Washington, D.C.: National Academy Press. 79-81.

Gribben, John. "Global Warming is Linked to Sahel Drought." *New Scientist.* April 24, 1986. 24.

Halacy, D. S., Jr., 1978. *Ice or Fire? Surviving Climatic Change.* New York: Harper & Row.

Hamilton, A. C. 1982. *Environmental History of East Africa: A Study of the Quaternary.* New York: Academic Press.

Lockwood, John G. 1985. *World Climatic Systems.* London: Edward Arnold.

Lurie, Edward. 1960. *Louis Agassiz: A Life in Science.* The University of Chicago Press.

Schultz, Gwen. 1963. *Glaciers and the Ice Age: Earth and its Inhabitants During the Pleistocene.* New York: Holt, Rinehart & Winston.

"The Very Air." *The Economist.* May 16, 1987. 92-93.

Weisburd, Stefi. "Polar-Equatorial Climate Link Reported." *Science News.* June 14, 1986.

▶ Grand Canyon, Arizona. The uplifting of the Colorado Plateau transformed a sluggish, lowland river into a fast-moving erosive powerhouse. Over about 15 million years, the Colorado River has cut its way through more than a mile and a half of rock and two billion years of Earth's history.

▶▶ The glowing reds of the Grand Canyon are a sign of oxygen, which reacts with iron on the surface of the rock just as rust coats a nail.

◀◀ The accelerating destruction of tropical rain forests is more than an aesthetic loss. Containing some 80 per cent of all the plant species on Earth, rain forests are an irreplaceable storehouse of genetic diversity.

THE HOME PLANET

Beard, Tim. "Ozone Watch." *Scientific American.* November 1987.

Begley, Sharon, and Gerald C. Lebenow. "Gushers at 30,000 Feet." *Newsweek.* June 27, 1988. 53.

Bjerklie, David. "Cloudy Crystal Balls." *Time.* October 19, 1987. 64.

"Brittle Plants Point to Ozone Damage." *Science News.* July 9, 1988. 28.

Carter, Vernon Gill and Tom Dale. 1974. *Topsoil and Civilization.* Norman: University of Oklahoma Press.

Foss, Clive. 1979. *Ephesus After Antiquity: A Late Antique, Byzantine and Turkish City.* Cambridge University Press.

Freundlich, Naomi. "Biosphere." *Popular Science.* December 1986. 54-56.

"Golf Courses May Cool the Desert." *Science News.* September 24, 1988. 203.

"Huge Ice Cube In Antarctic Waters." *Science News.* November 21, 1987. 326.

Johnson, Paul. 1978. *The Civilization of Ancient Egypt.* Ch. 1. New York: Atheneum.

Kramer, Mark. 1980. *Three Farms.* Boston: Little, Brown.

Kerr, Richard A. "The Carbon Cycle and Climate Warning." *Science.* December 6, 1983. 1107.

Lamb, H. H. 1982. *Climate, History and the Modern World.* London: Methuen.

Lehmonick, Michael D. "The Heat is On." *Time.* October 19, 1987. 58-67.

Lhote, Henri. "Oasis of Art in the Sahara." *National Geographic.* August 1987. 181-185.

Meiggs, Russell. 1982. *Trees and Timber in the Ancient Mediterranean World.* Oxford: Clarendon Press.

Mlot, C. "Trickle-Down Effects of Carbon Dioxide Rise." *Science News.* November 17, 1986. 309.

Monastersky, Richard. "New Chemical Model, New Ozone Fear." *Science News.* September 3, 1988. 148.

Monastersky, Richard. "Rising Sea Levels: Predictions and Plans." *Science News.* November 21, 1987. 326.

Monastersky, Richard. "Scientist Says Greenhouse Warming Is Here." *Science News.* July 12, 1988. 4.

Raloff, Janet. "New Research Clouds Pollution Picture." *Science News.* September 17, 1988. 180.

Raloff, Janet. "Pacific's CO_2 Levels: Cause For Concern?" *Science News.* March 24, 1986. 129.

Romer, John. 1982. *People of the Nile: Everyday Life in Ancient Egypt.* New York: Crown.

Shell, Ellen Ruppel. "Solo Flights Into Ozone Hole Reveal Its Causes." *Smithsonian.* February 1988. 142-155.

Thirgood, J. V. 1981. *Man and the Mediterranean Forest.* Ch. 1. New York: Academic Press.

ACKNOWLEDGMENTS

The authors wish to thank the scientists on both *The Miracle Planet* television and book advisory panels who patiently answered questions and corrected misapprehensions. The authors take full responsibility for any errors remaining. Thanks also to Charles Raymond and Sarah Hoffman of the University of Washington.

This book would not have been possible without Simon Griffith and Guy-Luc Levesque, indefatigable researchers at KCTS Television, and Barry Stoner and Katie Jennings, producers of *The Miracle Planet* television series.

The authors are especially grateful to Midori Kase for her translations of hundreds of photograph captions from the Japanese text.

Thanks also to Sue Burk for the initial book design; Tatsuko Nagasawa and Japan UNI Agency Inc., Teresa Fallace and NASA Representative in Australia, Mike Gentry and NASA Lyndon B. Johnson Space Center, Texas, and Harald Sund of Seattle for the supply of photographs.

◀ The rocky highlands of the Sahara are slowly eroding into sand. Despite the arid climate, the rocks' most effective sculptor is water.

◀◀ Irrigated land along the Columbia River contrasts with the barren plateau above. Photograph: Harald Sund.

▶▶ In contrast to the giant mechanized irrigation of the North American plains, Nepali farmers near Pokhara have laboriously terraced their steep hillsides to conserve water and lessen erosion.

INDEX

GEOLOGICAL TIME SCALE

Era	Period (million years ago)	Epoch	Evolutionary events
Cenozoic	Quaternary	Recent (0.01 -present) Pleistocene (2-0.01)	Rise of civilizations increase in number of herbs First *Homo*; height of last ice advance
	Tertiary	Pliocene (7-2) Miocene (26-7) Oligocene (38-26) Eocene (54-38) Paleocene (65-54)	First humans Dominance of land by angiosperms Dominance of land by mammals, birds, and insects
Mesozoic	Cretaceous (145-65)		K–T boundary; collision of India and Asia; last of the dinosaurs; second great dispersal of insects; angiosperms arise and expand as gymnosperms decline
	Jurassic (200-145)		Collision forming Manicouagan; dinosaurs abundant; first birds; gymnosperms (esp. cycads and conifers) still dominant; last of the seed ferns
	Triassic (245-200)		First mammals Dominance of land by gymnosperms; further decline of lycopsids and sphenopsids First dinosaurs
Paleozoic	Permian (285-245)		Extinction c. 96% of all species; formation of Pangaea; great expansion of reptiles; decline of amphibians; last of the trilobites; precipitous decline of lycopsids, sphenopsids, and seed ferns
	Carboniferous (360-285)		Great coal forests, dominated at first by lycopsids and sphenopsids, and later also by ferns and gymnosperms; age of Amphibians; first reptiles; first great dispersal of insects
	Devonian (430-360)		Expansion of primitive vascular plants, origin of first seed plants (gymnosperms) toward end of period; first liverworts; age of Fishes; first amphibians and insects
	Silurian (500-430)		Invasion of land by the first vascular plants toward end of period; remains of Cooksonia (first known vascular plant) found in Silurian sediments; invasion of land by a few arthropods
	Ordovician (570-500)		Marine algae abundant; first vertebrates (Agnatha)
	Cambrian (700-570)		Primitive marine algae (esp. Cyanophyta and probably Chlorophyta); stromatolite-building cyanobacteria; marine invertebrates abundant (including representatives of most phyla)
Precambrian	(3800-700)		Primitive marine life
Azoic	(4600-3800)